Chinese Military Manual Series
No. 1

# Militia Marksmanship Training Manual (Rifles and Submachine Guns)

Type 53 Mosin Nagant, Chiang Kai-shek Mauser, Arisaka and M1903 Springfield Rifles; Type 50 (PPSh-41) and Type 54 (PPS-43) Submachine Guns

Translated Chinese Military Manual featuring colour photographs and descriptions of Militia weapons and accessories for the collector

Edwin H. Lowe

## About The Author

Edwin H. Lowe has been a university researcher and academic for over 20 years in the diverse fields of conservation genetics, molecular biology, defence and strategic studies, Chinese history, Chinese politics and Chinese to English translation. He has worked and taught at Macquarie University, the Australian National University, the University of New South Wales and the Centre for Defence and Strategic Studies, Australian Defence College. Edwin has authored and co-authored articles and books in these fields, and has edited translations of Chinese texts. He is a writer and publisher, and this is the first in a series of translations of historical Chinese military manuals. Edwin H. Lowe was most recently a lecturer in Chinese studies and now a Visiting Associate in the Faculty of Arts at Macquarie University.

## About This Series

The Chinese Military Manual Series are translations of historical Chinese military manuals issued for the training of the People's Liberation Army and the Militia. First published in the People's Republic of China, these are the first publically available translations of Chinese military manuals in English.

## About This Book

This book is an illustrated training manual first published in 1973 for the training of the Chinese Militia in World War II era weapons. These are the Type 53 Mosin Nagant, Zhongzheng (Chiang Kai-shek) Mauser, Type 38 Arisaka, Type 99 Arisaka and M1903 Springfield rifles; the Type 50 (PPSh-41) and Type 54 (PPS-43) submachine guns. The manual is presented as a complete and original translation.

Collectors and shooters of the rifles and submachine guns described in this book will find the manual useful. It contains the official Chinese military training instructions for the maintenance, marksmanship and field employment of these weapons, as well as data for the various weapons and ammunition types.

Of particular interest to collectors, this book features colour photographs, descriptions and translations of markings of representative Chinese Militia weapons, accessories, ammunition labels, field equipment and accoutrements described in the manual.

## Further Information

The author would welcome any comments or further information about any topic raised in this book. Further information about the author and this series can be found at www.edwinhlowepublishing.com

# In This Series By The Author

*Militia Marksmanship Training Manual (Rifles and Submachine Guns).* Type 53 Mosin Nagant, Chiang Kai-shek Mauser, Arisaka and M1903 Springfield Rifles; Type 50 (PPSh-41) and Type 54 (PPS-43) Submachine Guns.

*People's Liberation Army Marksmanship Training Manual.* Type 56 (SKS) Carbine, Type 56 (AK-47) and Type 63 Assault Rifles; Type 56 (RPD) Light Machine Gun.

*People's Liberation Army Marksmanship Training Manual (Pistols).* Type 54 (Tokarev) and Type 59 (Makarov) Pistols.

*Militia Marksmanship Training Manual (Light Machine Guns).* Type 53 (DPM) & DP Degtyaryov, Brno ZB-26, Bren Mk II and Type 11 Nambu Light Machine Guns.

*People's Liberation Army Marksmanship Training Manual (Heavy Machine Guns).* Type 53 (Goryunov) and Type 58 (RP-46) Machine Guns.

*Militia Marksmanship Training Manual (Rifles, Light Machine Guns, Heavy Machine Guns)*


# Companion To This Series By The Author

*Everyone A Soldier! The Chinese Militia 1958 – 1984.*


# About Firearms Safety

While this book is a translation of the original Chinese military manual, it is intended solely to be a historical military document to be used for scholarly purposes. It is not intended to be an operator's manual for the modern collector or shooter of any of the firearms described in the book. The procedures and information in this book are those originally issued by the People's Liberation Army and they are neither endorsed nor recommended by the author and publisher.

Collectors and shooters should observe and comply with their local firearms laws in regard to safety, training, possession, handling and storage of firearms and ammunition.

The author and publisher will not be held responsible for any injury, death or damage incurred in the proper or improper storage, handing or use of any firearm or ammunition.

Chinese Military Manual Series
No. 1

# Militia Marksmanship Training Manual (Rifles and Submachine Guns)

Type 53 Mosin Nagant, Chiang Kai-shek Mauser, Arisaka and M1903 Springfield Rifles; Type 50 (PPSh-41) and Type 54 (PPS-43) Submachine Guns

Translated Chinese Military Manual featuring colour photographs and descriptions of Militia weapons and accessories for the collector

Edwin H. Lowe

www.edwinhlowepublishing.com

Translation and New Text
Copyright © Edwin H. Lowe 2015

All rights reserved. This book is copyright. Apart from any fair dealing for the purpose of private study, research, criticism or review, as permitted under the Copyright Act, no part may be reproduced or stored by any process without prior written permission of the publisher.

Grateful acknowledgement is made to copyright holders of materials reproduced in this book. Every reasonable effort has been made to trace the copyright holders. The publisher would be grateful to be informed of any inadvertent errors or omissions in the use of copyright material and would be pleased to correct them in any forthcoming editions.

Original text and illustrations published as *Minbing Sheji Jiaocai (Jiushi Buqiang, Chongfengqiang)* [ *Militia Marksmanship Training Manual (Obsolete Rifles and Submachine Guns)* ] by the People's Liberation Army, General Staff Headquarters, Mobilisation Department, March 1973.

**National Library of Australia Cataloguing-in-Publication entry**
Lowe, Edwin, 1972- author, translator.

Militia marksmanship training manual (rifles and submachine guns) : type 53 Mosin Nagant, Chiang Kai-Shek Mauser, Arisaka and M1903 Springfield rifles; type 50 (PPSH-41) and type 54 (PPS-43) submachine guns / Edwin H. Lowe author and translator.

9780994168207 (paperback)
Chinese military manual series ; no.1.
Military weapons--China--Handbooks, manuals, etc.
Military education--China--Handbooks, manuals, etc.
World War, 1939-1945--Equipment and supplies--Pictorial works
Military weapons--China--Pictorial works.
940.5425

Edwin H. Lowe Publishing, Sydney.
www.edwinhlowepublishing.com
contact@edwinhlowepublishing.com
Edwin H. Lowe Publishing is the trading name of Edwin Hulme Lowe. ABN 60 901 995 995

ISBN-13: 978-0-9941682-0-7
ISBN-10: 0994168209

Available from edwinhlowepublishing.com, Amazon.com and other book stores.
Printed by Createspace.

# Acknowledgements

The contributions of many people have made this book possible in its ultimate form.

I thank Shiyuan Chen for providing essential assistance in initiating both this book project and the Chinese Military Manual Series. I thank also James E. Wilen Jr. for sharing his collector's passion for the Chinese Type 53 carbine. Jamie's enthusiasm and passion for knowledge as a collector of the Type 53 carbine and accessories helped to inspire me to produce this book.

The following collectors kindly gave permission for the use of their photographs in this book: Richard Babb, Howard A. Bearse, Teri Jane Bryant, Roger Finzel, Stephen Fitchie, Bob Hanes, C.R. Holt, John Jett, Danny Nichols, Dan Pickle, Richard Simmons and James E. Wilen Jr. Howard A. Bearse, Bob Hanes, Dan Pickle and James E. Wilen Jr. in particular, generously sent me many photographs, of which only a few were used in the book.

Kenneth Elks, Bob Hanes, Tom Kulik, Bin Shih, Timothy Tietz and Lan Zhang all kindly shared with me information from their respective areas of expertise. My discussions with them helped to answer research questions that arose during the translation and writing of this book. I thank also the international community of collectors, particularly those contributing to the gunboards.com forum, for sharing their information, pictures and opinions with me over the years, which have helped to shape the wider body of collector's knowledge.

I also thank Alan Carvey, Chris Easton and Adrian Lowe for their technical and professional expertise, advice and comments on my translation and writing, and their many years of continued friendship.

My greatest thanks go to Emma Runcie for her unwavering support and faith.

# Table of Contents

**Introduction** ............................................................................................................ vii

**Militia Marksmanship Training Manual (Obsolete Rifles and Submachine Guns)** ............ I

    Contents .................................................................................................................. V

    General Description of Weapons ............................................................................... 1

        Rifles ................................................................................................................... 1

        Submachine Guns ............................................................................................. 15

    The Theory of Small Arms Fire ............................................................................... 23

    Determining Range .................................................................................................. 30

    Marksmanship ......................................................................................................... 32

        Rifles ................................................................................................................. 32

        Submachine Guns ............................................................................................. 42

    Marksmanship under Combat Conditions ............................................................... 50

    Appendices .............................................................................................................. 59

**Militia Weapons and Accessories for the Collector** ..................................................... 63

**Glossary** .................................................................................................................... 90

Militia Training. Ngari Prefecture, Tibet Autonomous Region, c.1970. Photo: Yuan Quan

# Introduction

This book is based on the training manual first published by the Mobilisation Department of the General Staff Headquarters, People's Liberation Army in 1973 as *Minbing Sheji Jiaocai (Jiushi Buqiang, Chongfengqiang)* [ *Militia Marksmanship Training Manual (Obsolete Rifles and Submachine Guns)* ].[1] The purpose of this manual was for the training of the Chinese Militia in 'obsolete' World War II era small arms still in service at that time. These were the Type 53 Mosin Nagant, Zhongzheng (Chiang Kai-shek) Mauser, Type 38 Arisaka, Type 99 Arisaka and M1903 Springfield rifles; the Type 50 (PPSh-41) and Type 54 (PPS-43) submachine guns.

This manual is presented as a complete and original translation. I have attempted to preserve the structure and the layout of the original manual as much as possible, and it is presented as a section within this book. The only significant difference between this translation and the original is that in a few places, I have appended the unfamiliar Chinese military designations of these weapons with the common names by which they are known in the West, eg 79 Rifle (Chiang Kai-shek Mauser).

This manual contains the official Chinese military training instructions for the maintenance, marksmanship and field employment of these weapons, as well as data for the various weapons and ammunition types. Collectors and shooters of rifles and submachine guns described in this book, as well as scholars of the Chinese military of the Cold War, will this manual useful. I believe that this book is the first English translation of any small arms manual of the People's Liberation Army (PLA) available to the public, and it represents a valuable insight into the training and doctrine of the PLA and the Militia at the time of its original publication.

The period during which the original manual was published in 1973 represents a time of reinvigoration of the Chinese military. Training and proficiency had declined from 1964 onwards, as the PLA sought to be "better red than expert", emphasising political education at the expense of professionalism and military capability. Additionally, like every other institution in the country, the social and political chaos of the Cultural Revolution beginning in 1966 wreaked havoc on the Militia, in part due to the competing interests inherent in its dual command structure, shared between the Communist Party of China and the People's Liberation Army. The opportunities afforded an armed force like the Militia in local Party power struggles saw Militia units embroiled in the political chaos that swept the country.

However the Sino-Soviet border conflict of 1969 raised the very real possibility of war with the Soviet Union and the spectre of the invasion of China. This resulted in an urgent renewal of training materials for the PLA and Militia. A range of new training manuals and posters ranging from small arms to anti-tank and anti-aircraft defences were published beginning in 1970 through to 1973. This was the first publication of training materials of this type since the beginning of the political radicalisation of the PLA in 1965, when the *Quotations of Chairman Mao Zedong* (aka 'The Little Red Book') became in essence, the primary 'training manual'. Prior to this, the last major period of publication of training manuals occurred in the period from 1958 to 1965, during the expansion of the Militia and the rapid introduction of new small arms such as the Type 56 (SKS) carbine, Type 56 (AK-47) assault rifle and Type 56 (RPD) light machine gun into the PLA.

This 1973 manual, *Militia Marksmanship Training Manual (Obsolete Rifles and Submachine Guns)* follows a 1972 manual *Militia Marksmanship Training Manual (Provisional Edition)*, itself a compilation incorporating an earlier 1963 manual, *Militia Rifle Marksmanship Training Manual*.[2] While it appears that the 1973 manual is based on the earlier 1963/1972 manuals, there are a number of differences between the two. The 1963 manual is focused on

---

[1] People's Liberation Army, General Staff Headquarters, Mobilisation Department, *Minbing Sheji Jiaocai (Jiushi Buqiang, Chongfengqiang)* [*Militia Marksmanship Training Manual (Obsolete Rifles and Submachine Guns)*] (Beijing: People's Liberation Army, 1973).
[2] People's Liberation Army, Nanjing Military Region Headquarters, *Minbing Sheji Jiaocai (Shiyongben)* [*Militia Marksmanship Training Manual (Provisional Edition)*], (Nanjing: People's Liberation Army, 1972); People's Liberation Army, General Staff Headquarters, Mobilisation Department, *Minbing Buqiang Sheji Jiaocai* [*Militia Rifle Marksmanship Training Manual*], (Beijing: People's Liberation Army, 1963).

five rifle types, in the following order of reference: the Zhongzheng (Chiang Kai-shek) Mauser, Type 53 Mosin Nagant carbine (with reference to the Mosin Nagant 91/30 rifle and M44 carbine), Type 38 Arisaka, Type 99 Arisaka and the M1903 Springfield. The Type 53 Mosin Nagant carbine, manufactured in China between 1953 and 1956, entered Militia service only after it had been phased out of the PLA, replaced by the Type 56 (SKS) carbine beginning in 1958 - 1959. The other rifles were World War II era weapons. The 91/30 and M44 Mosin Nagants were Korean War purchases from the Soviet Union, and the Arisakas were captured either from the Japanese, or from the Nationalist Government during the Chinese Civil War, along with Nationalist Mausers and Springfields. By 1973, the Type 53 carbine had become the primary bolt action rifle of the Militia and had overtaken the Mauser in importance. This change is reflected in the 1973 manual.

In general terms, the 1963 manual was more detailed in certain chapters. In the chapter on the theory of small arms fire, the concepts of trajectory, dead space, dangerous ground and defilade were explained more cogently and in much greater detail. The 1973 text in this chapter was wholly inadequate in places, as these appeared to be extracted from the 1963 manual without the complete context. Similarly, the 1963 manual included techniques or instructions which were not covered in the 1973 manual. These included the conduct of range practises, as well as descriptions of a wider range of training aids for marksmanship, such as a simple up-scaled wooden sighting device which simulated rifle sights; an aim checking technique based on bore sighting devices; and a much more detailed description of the 'Four Point Aim Check' technique. This latter subject was wholly inadequate in the 1973 manual and could only be understood and translated after a reading of the 1963 text. Any such shortcomings in the 1973 text were corrected or annotated for cogency in this translation. In comparison, the 1973 manual was less theoretical in nature than the 1963 manual, and much more practical. It included more detailed information on the field usage of the various weapons, such as practical marksmanship, firing in defensive positions, diverse terrains and air defence. It also omitted the chapter on grenades, which appeared in the 1963 manual. These differences reflect the urgency of Militia training as a result of the pressure of Sino-Soviet border tensions, as well as the deficiencies in Militia training as a result of the neglect of military skills during the Cultural Revolution.

The World War II era weapons in this manual were the mainstay of the Militia throughout the 1950s, including the beginning of the mass expansion of the Militia in 1958 with the 'Everyone A Soldier' movement, and into the early 1960s. However from 1960, the Militia began the process of re-equipping with standard PLA weapons and ammunition. An isolated production batch of Type 53 carbines produced in 1960, long enigmatic amongst collectors, was probably intended to meet Militia requirements.[3] Period photographs from the late 1960s and early 1970s show that the Type 53 carbine had become the primary Militia bolt action rifle, by virtue of its use of PLA standard ammunition, and was itself increasingly overshadowed by the Type 56 (SKS) carbine as the primary individual weapon. Similarly in the same period, the Type 50 and Type 54 submachine guns were increasingly being supplanted by the Type 56 (AK-47) assault rifle as the squad selective fire weapon. However by 1973 when this manual was published, these bolt action rifles and submachine guns had all been designated 'obsolete'. These 'obsolete' weapons stayed in service with the Militia as late as 1981, when the last of the obsolete weapons were withdrawn, along with the general confiscation of weapons still remaining in civilian hands following decades of war and the general militarisation of society. These obsolete Militia weapons were then sold on the international market and began to be acquired by Western collectors in the early 1980s.

When Western collectors encountered these Militia weapons, it was the first time that large numbers of these various weapons had been freely available to collectors. The condition in which these weapons were found ranged from excellent to very poor, and in most cases, reflected the many decades that these weapons had been actively used in service, not only by the Militia, but the PLA, the Nationalist or the Japanese armies before them. Many examples showed clear neglect or poor maintenance, while other examples reflected poor conditions of storage and transport after the end of their military service. The poor conditions of these

---

[3] Apart from the isolated 1960 production run, Type 53 Carbines were produced continuously from 1953-1956.

weapons, the damage and the missing or substituted parts such as cleaning rods, combined with the scarcity of information about them, resulted in a number of long standing myths about the quality or nature of these various weapons. This book includes a section which discusses these Militia weapons from the perspective of the collector and addresses these myths and misconceptions, drawing on the information provided by collectors in the years since the export of these weapons from China.

During the course of the long and varied service lives of these weapons, a range of interesting markings, both official and unofficial, have been applied to the rifles in particular, mirroring the dramatic sweep of Chinese history in the 20$^{th}$ Century. This book showcases a number of interesting Militia markings found on collector's rifles, and includes colour photographs, translations of the markings and a discussion on their organisational or historical importance. Through the understanding of the nature and condition of these Militia weapons and some of their distinctive markings, this section will help collectors to better appreciate these weapons in their full and rich historical context.

Similarly, this book includes photographs, translations of markings and descriptions of representative Chinese ammunition, weapons accessories, field equipment and Militia accoutrements, which have become available to the collector. This book showcases various collector's items and shows them in their service context as depicted in period photographs of the Militia. This section will provide collectors with a better appreciation of these weapons in Militia service. It will reveal the diversity and 'correctness' of the various combinations of Militia weapons and equipment, from the perspective of the collector.

For the benefit of readers who have an interest in the Chinese Militia, I have written a companion book to this 'Chinese Military Manual Series', *Everyone A Soldier! The Chinese Militia 1958-1984*. A more complete appreciation of the Chinese Militia may be gained from this book.

Edwin H. Lowe
Sydney, 2015.

Militia Training, c.1965

# Militia Marksmanship Training Manual (Obsolete Rifles and Submachine Guns)

People's Liberation Army,

General Staff Headquarters, Mobilisation Department,

March 1973.

# Quotation From Chairman Mao Zedong

The Militia is the basis of victory.

The whole Party must attach great importance to warfare by learning military skills and being prepared to go to war.

We should make use of the intervals between battles to stress the training of troops. This applies to field armies, regional troops and militia. As for the training courses, the main objective should still be to raise the level of technique in marksmanship, bayoneting, grenade-throwing and the like, and the secondary objective should be to raise the level of tactics, while special emphasis should be laid on night operations. As for the method of training, we should unfold the mass training movement in which officers teach soldiers, soldiers teach officers and the soldiers teach each other.[4]

---

[4] Translator's Note: From Mao Zedong, 'Policy For Work In The Liberated Areas For 1946', 15 December 1945; in *Selected Works of Mao Tse-tung* Vol 4. (Peking: Foreign Languages Press 1969), p.76. At the time of the original publication of this manual in 1973, China was in the midst of the socio-political movement known as the Cultural Revolution (1966-1976). During the Cultural Revolution, the personality cult of Mao Zedong was raised to extreme levels and the Chinese people were essentially required to read the writings of Mao, most famously from the *Quotations From Chairman Mao Zedong* ('The Little Red Book'). Quotations from Mao Zedong appeared highlighted in red at the beginning of all books, quoted in articles and broadcasts, and even going as far as appearing on facia plates of Chinese military equipment such as army radio sets. This quote from 1945 advocating the 'mass training movement' epitomised the institutionalised egalitarian nature of the PLA and Militia during the Maoist era, and the primacy of ideological purity over professional ability under the Maoist ideal of being "better red than expert".

# Explanation

In order to meet Militia training requirements, we have complied three marksmanship training manuals on Obsolete Rifles and Submachine Guns, Light Machine Guns and Heavy Machine Guns. If there is insufficient information, or if there are other problems resulting from the hasty compilation of these manuals, please advise us promptly.[5]

---

[5] Translator's Note: These three marksmanship training manuals published in 1973 followed an earlier 1972 single volume compilation: People's Liberation Army, Nanjing Military Region Headquarters, *Minbing Sheji Jiaocai (Shiyongben)* [*Militia Marksmanship Training Manual (Provisional Edition)*], (Nanjing: People's Liberation Army, 1972).

# Contents

General Description of Weapons ............................................................................................ 1
Rifles ............................................................................................................................................ 1
   1. Combat Capabilities ........................................................................................................... 1
   2. Description of Parts and Function, Disassembly and Assembly ................................. 1
      Type 53 Carbine (Mosin Nagant) ...................................................................................... 1
         A. Description of Parts and Function ......................................................................... 1
         B. Disassembly and Assembly ..................................................................................... 4
            1. General Notes ..................................................................................................... 4
            2. Disassembly ....................................................................................................... 4
            3. Assembly ............................................................................................................ 5
      79 Rifle (Chiang Kai-shek Mauser) .................................................................................. 6
         A. Description of Parts and Function ......................................................................... 6
         B. Disassembly and Assembly ..................................................................................... 7
            1. General Notes ..................................................................................................... 7
            2. Disassembly ....................................................................................................... 7
            3. Assembly ............................................................................................................ 8
      65 Rifle (Type 38 Arisaka) ................................................................................................. 9
         A. Description of Parts and Function ......................................................................... 9
         B. Disassembly and Assembly ................................................................................... 10
            1. General Notes ................................................................................................... 10
            2. Disassembly ..................................................................................................... 10
            3. Assembly .......................................................................................................... 10
      Type 99 Rifle (Type 99 Arisaka) ..................................................................................... 11
      30 Rifle (M1903 Springfield) .......................................................................................... 11
         A. Description of Parts and Function ....................................................................... 11
         B. Disassembly and Assembly ................................................................................... 12
            1. General Notes ................................................................................................... 12
            2. Disassembly ..................................................................................................... 12
            3. Assembly .......................................................................................................... 13
   3. Weapons Care and Stoppages / Malfunctions ............................................................. 13
      A. Weapons Care ................................................................................................................ 13
      B. Stoppages / Malfunctions ............................................................................................. 14
         Table: Stoppages / Malfunctions, Causes and Corrective Actions ............................ 14

## Submachine Guns .................................................................................................................................. 15
### 1. Combat Capabilities ..................................................................................................................... 15
### 2. Description of Parts and Function, Disassembly and Assembly ............................................ 15
#### Type 50 Submachine Gun (PPSh-41) ........................................................................................ 15
##### A. Description of Parts and Function ................................................................................. 15
##### B. Disassembly and Assembly .......................................................................................... 17
###### 1. General Notes ........................................................................................................ 17
###### 2. Disassembly ........................................................................................................... 17
###### 3. Assembly ............................................................................................................... 18
#### Type 54 Submachine Gun (PPS-43) ........................................................................................... 18
##### A. Description of Parts and Function ................................................................................. 19
##### B. Disassembly and Assembly .......................................................................................... 20
###### 1. General Notes ........................................................................................................ 20
###### 2. Disassembly ........................................................................................................... 20
###### 3. Assembly ............................................................................................................... 21
### 3. Weapons Care and Stoppages / Malfunctions ........................................................................ 22
#### A. Weapons Care ....................................................................................................................... 22
#### B. Stoppages / Malfunctions .................................................................................................... 22
##### Table: Stoppages / Malfunctions, Causes and Corrective Actions ................................. 22

## The Theory of Small Arms Fire ....................................................................................................... 23
### 1. Small Arms Fire and Recoil ........................................................................................................ 23
#### A. Small Arms Fire .................................................................................................................... 23
#### B. Recoil ..................................................................................................................................... 24
### 2. Practical Implications of Trajectory ......................................................................................... 24
#### A. Trajectory .............................................................................................................................. 24
#### B. Dangerous Space .................................................................................................................. 25
#### C. Defilade and Dead Ground ................................................................................................. 25
### 3. Sighting and Aiming ................................................................................................................... 26
#### A. Using the Sights ................................................................................................................... 26
#### B. Sight Radius, Line of Sight, Point of Aim and Height of Trajectory .............................. 27
#### C. Setting the Sights and Selecting the Point of Aim ............................................................ 27
#### D. Observing the Shot and Correcting the Point of Aim ..................................................... 27

- 4. The Effect of External Factors on the Shot and the Point of Aim .................................................. 28
  - A. The Effect of Wind on the Shot ............................................................................................. 28
    - 1. Determining Wind Direction and Strength .................................................................... 28
    - 2. Allowing for Wind and Correcting the Point of Aim ..................................................... 28
      - Table: Correction for a Moderate Cross Wind ........................................................ 28
  - B. The Effect of Sunlight on the Point of Aim ........................................................................... 29

# Determining Range ............................................................................................................................. 30
- 1. Determining Range by Pace Counting ............................................................................................ 30
- 2. Determining Range by Eye ............................................................................................................. 30
  - A. Determining Range by Comparing Distances ....................................................................... 30
  - B. Determining Range by Degrees of Visibility ......................................................................... 30
  - C. Determining Range by Parallax ............................................................................................. 31

# Marksmanship ..................................................................................................................................... 32
# Rifles .................................................................................................................................................... 32
- 1. Safety Inspection ............................................................................................................................. 32
- 2. Loading and Sight Setting ............................................................................................................... 32
  - A. Prone Position ........................................................................................................................ 32
  - B. Kneeling Position ................................................................................................................... 33
  - C. Standing Position ................................................................................................................... 34
- 3. Marksmanship in Supported Positions ............................................................................................ 34
  - A. Prone Supported Position ...................................................................................................... 34
    - 1. Holding ............................................................................................................................ 34
    - 2. Aiming ............................................................................................................................. 36
    - 3. Firing ............................................................................................................................... 37
  - B. Kneeling Supported Position ................................................................................................. 37
  - C. Standing Supported Position ................................................................................................. 38
- 4. Marksmanship in Unsupported Positions ....................................................................................... 38
  - A. Prone Unsupported Position .................................................................................................. 38
  - B. Kneeling Unsupported Position ............................................................................................. 39
  - C. Standing Unsupported Position ............................................................................................. 39
- 5. Correcting the Aim .......................................................................................................................... 40
  - A. Individual Check .................................................................................................................... 40
  - B. Four Point Aim Check ........................................................................................................... 40
  - C. Aim Corrector ........................................................................................................................ 41

## Submachine Guns .................................................................................................................. 42
### 1. Safety Inspection ............................................................................................................. 42
### 2. Loading and Sight Setting .................................................................................................. 42
#### A. Filling Magazines ......................................................................................................... 42
#### B. Prone Position ............................................................................................................. 43
#### C. Kneeling Position ......................................................................................................... 43
#### D. Standing Position ......................................................................................................... 44
### 3. Marksmanship in Supported Positions ................................................................................ 44
#### A. Prone Supported Position ............................................................................................ 44
##### 1. Holding .................................................................................................................... 44
##### 2. Aiming .................................................................................................................... 45
##### 3. Firing ...................................................................................................................... 46
#### B. Kneeling Supported Position ........................................................................................ 47
#### C. Standing Supported Position ........................................................................................ 47
### 4. Marksmanship in Unsupported Positions ............................................................................ 48
#### A. Prone Unsupported Position ........................................................................................ 48
#### B. Kneeling Unsupported Position .................................................................................... 48
#### C. Standing Unsupported Position .................................................................................... 49

## Marksmanship under Combat Conditions .................................................................................. 50
### 1. Firing at Fleeting Targets ................................................................................................... 50
### 2. Firing at Moving Ground Targets ....................................................................................... 50
#### A. Crossing Target Allowance .......................................................................................... 50
##### Table: Moving Target Speeds (m/s) ............................................................................. 50
##### Table: Bullet Flight Time (s) ......................................................................................... 50
##### Table: Rifle - Crossing Target Allowance for Jogging Infantry (Body Widths) ................ 51
##### Table: Submachine Gun - Crossing Target Allowance for Jogging Infantry (Body Widths) ................ 51
#### B. Engaging Crossing Targets .......................................................................................... 51
##### 1. Ambush Shot ........................................................................................................... 51
##### 2. Tracking Shot .......................................................................................................... 51
### 3. Firing in Mountainous Terrain ............................................................................................. 52
##### Table: Allowance for Range and Angle in Mountainous Terrain - Point of Aim (cm) ......... 52

- 4. Firing at Airborne Targets ..................................................................................................... 52
  - A. Firing at Aircraft ............................................................................................................ 52
    - Table: Allowance for Firing at Aircraft with Rifles (Aircraft Lengths) ................... 54
  - B. Firing at Paratroopers ..................................................................................................... 55
    - Table: Allowance for Firing at Descending Paratroopers (Body Lengths) ............. 55
  - C. Positions for Firing at Airborne Targets ........................................................................ 56
- 5. Firing at Seaborne Targets ........................................................................................................ 57
- 6. Firing at Night .......................................................................................................................... 57
  - A. Firing in Flare Illumination ............................................................................................ 57
  - B. Firing in Moonlight ........................................................................................................ 57
  - C. Firing in Fixed Lines ...................................................................................................... 58

# Appendices ...................................................................................................................................... 59

- 1. Rifle Data .................................................................................................................................. 59
- 2. Submachine Gun Data ............................................................................................................. 59
- 3. Rifles - Height of Point of Impact above the Point of Aim .................................................... 60
- 4. Submachine Guns - Height of Point of Impact above the Point of Aim ............................... 60
- 5. Rifle - Aiming Off to Correct the Point of Impact ................................................................. 61
- 6. Ammunition Data ..................................................................................................................... 61
  - A. Type 53 Carbine (Mosin Nagant) Cartridges ................................................................ 61
  - B. 79 Rifle (Mauser) Cartridges .......................................................................................... 61
  - C. 65 Rifle (Arisaka) Cartridges .......................................................................................... 62
  - D. Type 99 Rifle (Arisaka) Cartridges ................................................................................ 62
  - E. 30 Rifle (Springfield) Cartridges .................................................................................... 62

# General Description of Weapons

# Rifles

## 1. Combat Capabilities

The rifle is the primary individual weapon of the Militia. Rifles fire with a high muzzle velocity, flat trajectory and high lethality. They are simple in design, light weight, easy to carry and simple to use. Rifles are most effective at firing at individual targets at ranges up to 400m. Concentrated fire can be used against aircraft and paratroopers at ranges up to 500m and against group targets up to 800m. The bayonet and rifle butt can also be used to kill the enemy in close quarter battle.

## 2. Description of Parts and Function, Disassembly and Assembly

### Type 53 Carbine (Mosin Nagant)

The Type 53 Carbine was first manufactured in China in 1953. The calibre is 7.62mm and it is designated the Type 53 Carbine (Fig.1).

Fig.1 Type 53 Carbine

**A. Description of Parts and Function**

The components of the carbine include the barrel, sights, receiver, bolt, trigger mechanism, magazine, stock, bayonet and accessories kit.

**1. Barrel.** (Fig.2) The barrel directs the flight of the bullet. The interior of the barrel consists of the rifled bore and the chamber. The chamber houses the cartridge and the rifling imparts spin on the bullet to provide accuracy and penetration. The rifling of the bore has four grooves and the distance between two opposite lands is the calibre of the carbine.

Fig.2 Rifle Barrel

**2. Sights.** (Fig.3) The sights are used for aiming the carbine. The sights consist of the backsight and foresight. Marked on the upper surface of the backsight leaf are graduations ranging from 1 – 10, with each graduation indicating 100m in range.

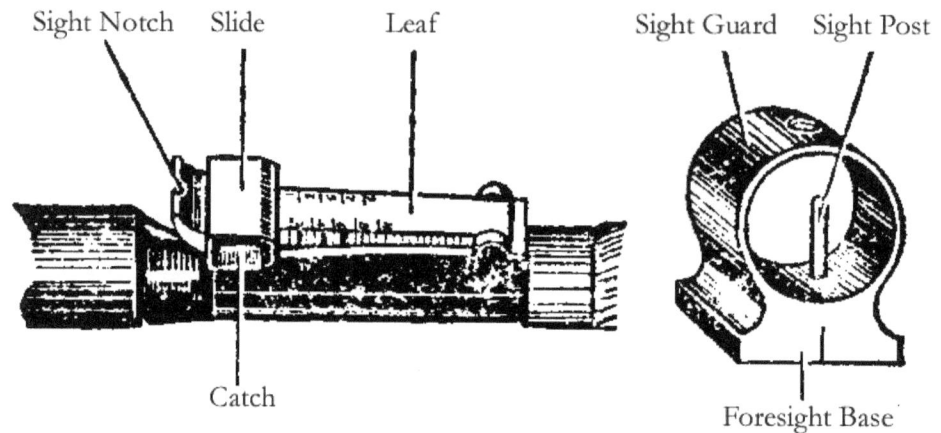

Fig.3 Sights

**3. Receiver.** The receiver houses the bolt and the assemblies of the operating mechanism. The ejector is located inside the receiver.

**4. Bolt Assembly.** (Fig.4) The bolt assembly is used for feeding the cartridge into the chamber, closing the breech, firing the shot and extracting the cartridge case. The bolt assembly consists of the bolt body, bolt head, connecting bar, striker, striker spring, cocking piece and extractor. The cocking piece also serves as the safety mechanism.

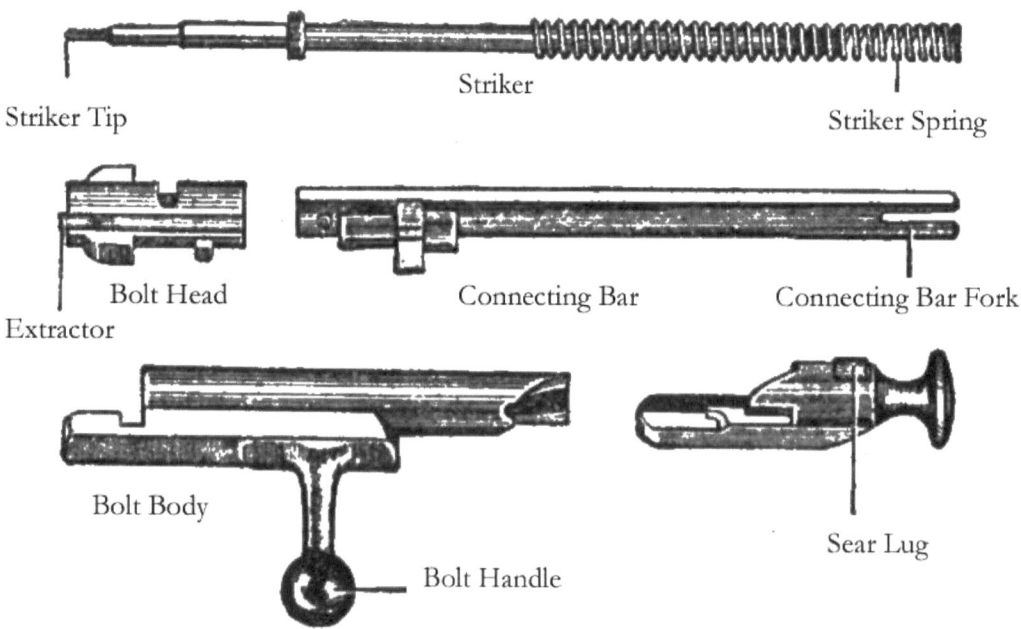

Fig.4 Bolt Assembly

**5. Trigger Mechanism.** (Fig.5) The trigger mechanism is used for firing the carbine. The trigger mechanism consists of the trigger, trigger spring, sear, bolt stop, pin and screw.

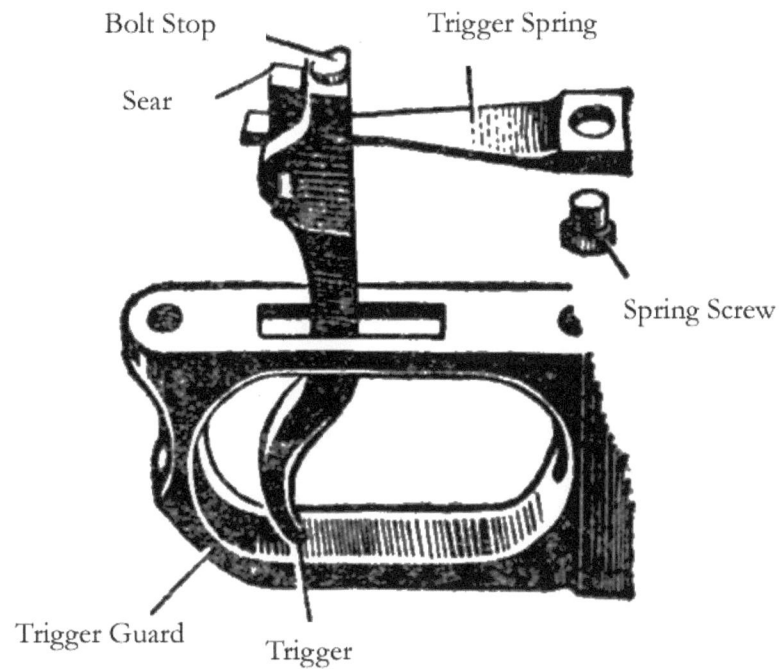

Fig.5 Trigger Mechanism

**6. Magazine Assembly.** (Fig.6) The magazine holds the cartridges. The magazine assembly consists of the magazine platform, spring, floorplate and trigger guard.

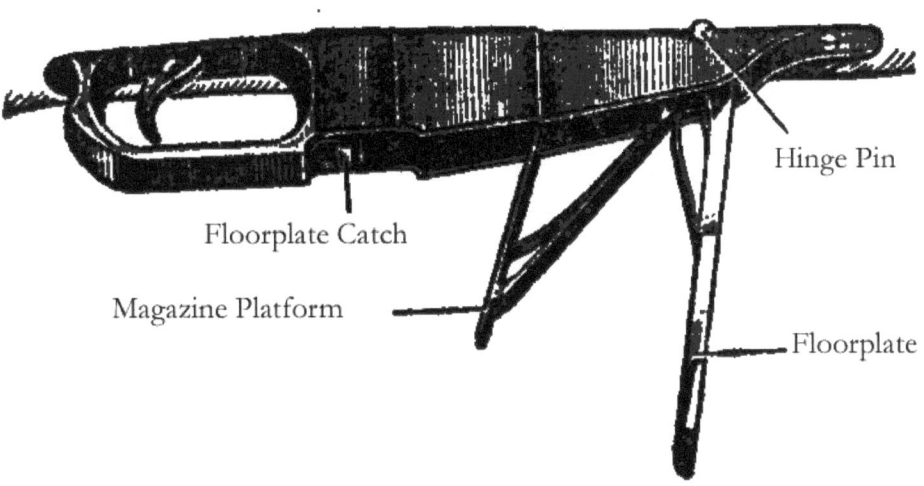

Fig.6 Magazine Assembly

**7. Stock.** The stock consists of three parts, the upper hand guard, fore-end and butt. The stock serves to support the barrel, and is used for carrying and operating the carbine.

**8. Bayonet and Accessories Kit.** The folding bayonet is used for killing the enemy in close quarter battle. The carbine's accessories kit consists of the cleaning rod, cleaning jag, muzzle guide, cleaning rod collar, drift punch, bolt tool, bore brush, oil bottle, sling and ammunition pouches.

## B. Disassembly and Assembly

### 1. General Notes

(1) Ensure that the carbine is unloaded prior to disassembly, in order to prevent accidental discharge.

(2) It is essential that disassembly and assembly of the carbine is conducted carefully, in order to prevent the loss or damage of parts.

(3) Disassembly and assembly of the carbine should be conducted on a clean surface.

### 2. Disassembly

(1) **Remove the Bolt.** Depress the trigger with the index finger of the left hand. Open the bolt with the right hand and pull the bolt to the rear and out of the receiver.

(2) **Fold the Bayonet.** Hold the carbine upright with the upper side of the carbine facing towards you. Hold the barrel with the left hand and pull the bayonet tube upwards with the right hand, until it clears the muzzle. Fold the bayonet backwards.

(3) **Remove the Cleaning Rod.** Grip the cleaning rod between the thumb and fingers of the right hand, twist to the left to unscrew the rod and pull it out.

(4) **Remove the Magazine Floorplate.** Depress the magazine floorplate catch with the right index finger and open the floorplate. Compress the magazine platform and spring, and remove the floorplate.

(5) **Disassemble the Bolt.** Hold the bolt in the left hand, grasping the bolt handle with the thumb and the bolt head with the index finger. With the right hand draw back the cocking piece far enough so that the cocking piece cam comes out of the bolt body cam groove, but not so far that the cocking piece sear lug comes out of the connecting bar fork. Rotate the cocking piece to the left and release it (Fig.7).

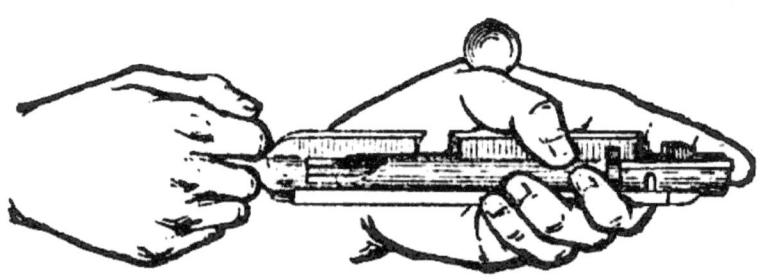

Fig.7 Disassemble the Bolt

Hold the cocking piece in the right hand and push forward the connecting bar and bolt head with the left hand. Remove the bolt head from the connecting bar.

Disassemble the cocking piece. Hold the bolt body vertically and rest the striker tip against a wooden surface. With the left hand, press down on the bolt handle as firmly as possible to compress the striker spring. With the right hand, rotate the cocking piece to the left to unscrew it from the striker and slowly release pressure from the bolt handle. Remove the striker and striker spring (Fig.8).

### 3. Assembly

(1) **Assemble the Bolt.** Insert the striker into the striker spring and then insert the striker assembly into the bolt body. Hold the bolt body vertically and rest the striker tip on a wooden surface. With the left hand, press down on the bolt handle and screw the cocking piece onto the striker with the right hand. Slowly release pressure on the bolt handle and ease the cocking piece cam into position with the bolt body cam groove. Use the large notch in the bolt tool to rotate the striker until the end of the striker is flush with the cocking piece and the index line of the striker is aligned with the index line on the rear of the cocking piece.

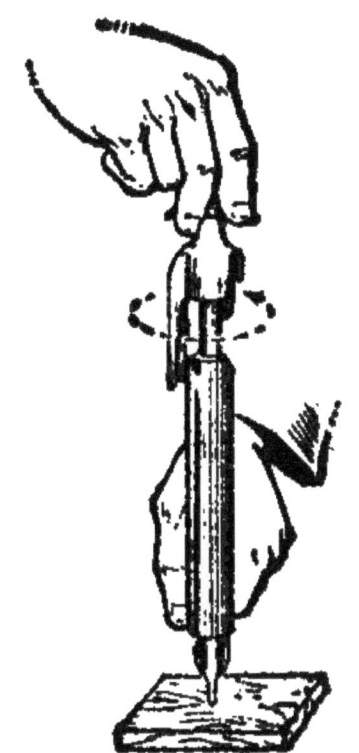

Fig.8 Disassemble the Cocking Piece

With the left hand attach the bolt head to the connecting bar. Rotate it to the right to a stop. With the right hand insert the striker into the connecting bar sleeve. Align the bolt head lug with the slot in the bolt body rib. Check the striker protrusion using notches on the bolt tool (Fig.9). The striker should pass the third notch ('95'), but not pass the second notch ('75'). Inspect the extractor and sear for wear or damage and that the connecting bar is not bent.

Hold the bolt handle with the left hand, with the index and middle fingers on the bolt head. Pull back the cocking piece with the right hand, and rotate it to the right, so that the cam on the cocking piece enters the notch on the bolt body.

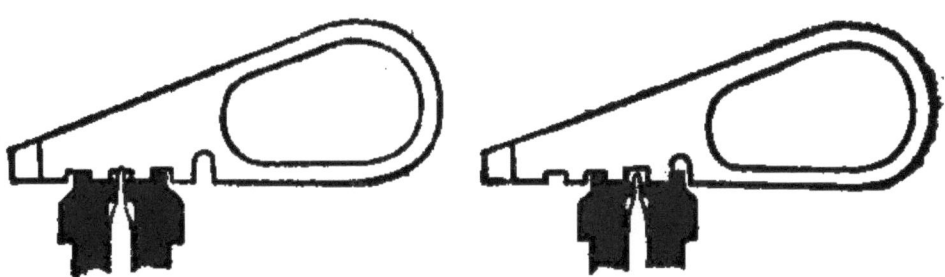

Fig.9 Check the Striker Protrusion

(2) **Attach the Magazine Floorplate.** Hold the carbine with the left hand and compress the floorplate, spring and magazine platform with the right hand. Insert the hook of the floorplate assembly onto the magazine hinge pin. Release the floorplate assembly and close the floorplate.

(3) **Fit the Cleaning Rod.** Insert the cleaning rod into the cleaning rod well and screw it to the right to tighten.

(4) **Extend the Bayonet.** Hold the carbine upright by the barrel with the left hand, and pull down on the bayonet with the right hand. Rotate the bayonet upwards fully to the right until the bayonet muzzle ring fits over the muzzle.

(5) **Fit the Bolt.** Hold the handgrip of the butt stock with the left hand, depress the trigger and insert the bolt into the receiver.

At present, there is another type of Soviet rifle used by the Militia. Its combat capabilities and features are essentially the same as those of the Type 53 Carbine. The differences are: the rifle weighs 4.5kg and is 166cm long with the bayonet fixed; the carbine weighs 3.9kg and is 133cm long with the bayonet extended. The rifle bayonet is longer and can be unfixed, whereas the carbine has a folding bayonet.

## 79 Rifle (Chiang Kai-shek Mauser)

The Type 24 Rifle is an improved version of the German Standard Modell Mauser rifle. The calibre is 7.9mm and it is designated the 79 Rifle (Fig.10).[6]

Fig.10  79 Rifle

### A. Description of Parts and Function

The components of the rifle include the barrel, sights, receiver, bolt (Fig.11), trigger mechanism, magazine, stock and bayonet. Their functions are the same as those of the Type 53 Carbine. The differences are:

1. The sight leaf is marked with graduations from 1 - 20. Each graduation indicates 100m in range.

2. The ejector and bolt stop are on the left hand side of the receiver.

3. The safety catch is on the top of the bolt sleeve and it is used to apply the safety mechanism.

---

[6] Translator's Note: The Chinese designation 'Type 24' specifically refers to the licensed Chinese copy of the Standard Modell Mauser short rifle. First produced in 1935, or the 24th Year of the Republic, it was initially designated the 'Type 24' and later the 'Type Zhongzheng', after the formal name of the Chinese Nationalist leader, Generalissimo Chiang Kai-shek (Jiang Zhongzheng). The 1963 Militia manual noted that the 'Type 24' was also known as the 'Type Zhongzheng', and that it had been introduced in 1935 by the 'Chiang Kai-shek bandits'. By the time that this manual was published in 1973 in the midst of the Cultural Revolution, the mention of the 'Chiang Kai-shek bandits' and of the 'Type Zhongzheng' in a PLA manual was politically unacceptable. The name '79 Rifle' was a common designation for all 7.92mm Mauser short rifles, including the Chinese made Zhongzheng and imported rifles such as the FN 1924/30, vz.24, Standard Modell, Kar98k etc, that remained in Communist service after the end of the Chinese Civil War. The Type Zhongzheng rifle is also known to collectors as the 'Chiang Kai-shek Mauser', or the 'Generalissimo rifle'.

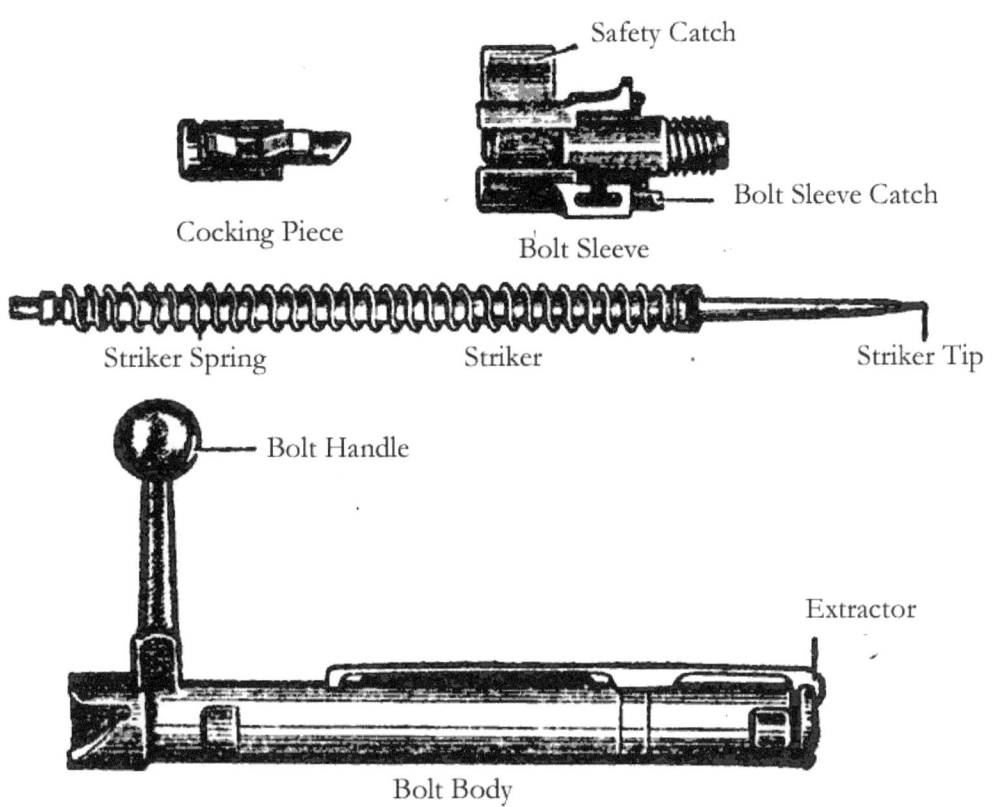

Fig.11 Bolt Assembly

## B. Disassembly and Assembly

### 1. General Notes
As for the Type 53 Carbine.

### 2. Disassembly
(1) **Remove the Cleaning Rod.** Hold the rifle upright, grip the cleaning rod between the thumb and fingers of the right hand, twist it to the left to unscrew the rod and pull it out.

(2) **Remove the Bolt.** Hold the rifle face up. With the right hand, lift the bolt handle upwards and close it down again. Lift the safety catch to the vertical position. Pull the bolt stop lever against the spring pressure with the thumb of the left hand. With the right hand, lift up the bolt handle and pull it to the rear and remove the bolt.

(3) **Remove the Magazine Floorplate.** Hold the rifle upside down with the left hand, with the thumb covering the magazine floorplate. With the right hand, insert the end of the cleaning rod into the circular hole in the floorplate and depress the spring loaded floorplate catch. Push the floorplate to the rear to release it from the trigger guard.

(4) **Disassemble the Bolt.** Hold the bolt body with the left hand. Hold the bolt sleeve with the right hand and depress the bolt sleeve catch. Unscrew the bolt sleeve to the left and separate it from the bolt body (Fig.12).

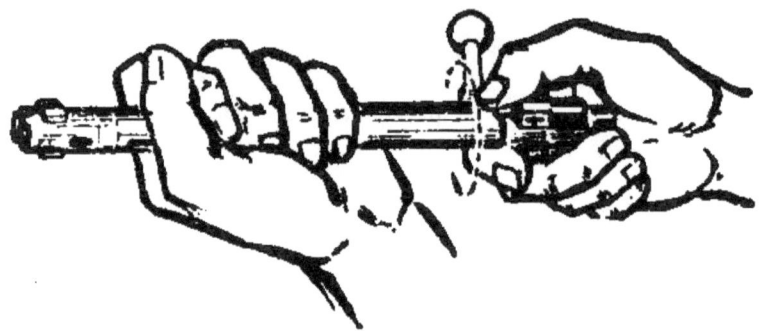

Fig.12 Disassemble the Bolt

Hold the bolt sleeve with the right hand and press down on the back of the safety catch with the thumb. Hold the striker vertically with the striker tip against a wooden surface and press down firmly to compress the striker spring. With the left hand, rotate the cocking piece 90 degrees and remove it. Slowly release pressure on the striker spring and remove the safety catch, striker and striker spring (Fig.13).

## 3. Assembly

(1) **Assemble the Bolt.** Insert the striker into the striker spring. Insert the safety catch into the bolt sleeve. Hold the striker vertically with the striker tip against a wooden surface and insert the striker into the bolt sleeve. With the thumb of the right hand on the safety catch, press down on the bolt sleeve as far as possible to compress the striker spring. With the left hand, attach the cocking piece to the rear of the striker and rotate it 90 degrees and slowly release the pressure on the striker spring until the bolt sleeve locks into position. Hold the bolt body with the left hand and insert the striker assembly into the bolt body. With the right hand, rotate the bolt sleeve to the right until it locks into position with the bolt body.

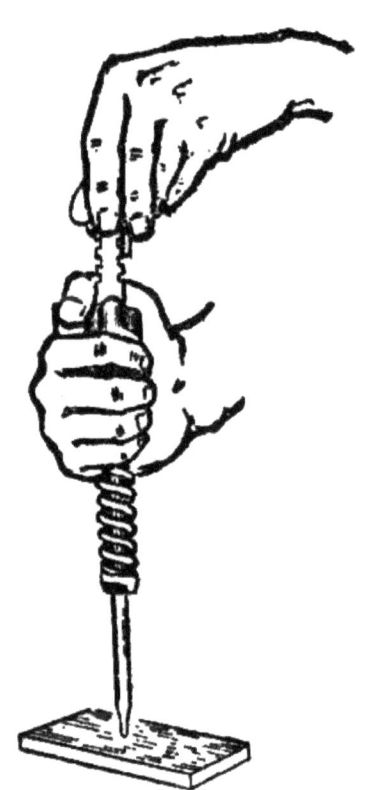

Fig.13 Disassemble the Bolt Sleeve

(2) **Attach the Magazine Floorplate.** Hold the rifle upside down and insert the magazine platform and spring into the magazine. Push the floorplate forward until the floorplate catch locks.

(3) **Fit the Bolt.** Hold the grip of the butt stock with the left hand and depress the trigger with the index finger. Insert the bolt with the right hand, close the bolt and push the safety catch to the left.

(4) **Fit the Cleaning Rod.** Insert the cleaning rod into the cleaning rod well and screw it to the right to tighten.

# 65 Rifle (Type 38 Arisaka)

The Type 38 Rifle was manufactured in Japan. The calibre is 6.5mm and it is designated the 65 Rifle (Fig.14).

Fig.14  65 Rifle

## A. Description of Parts and Function

The components of the rifle include the barrel, sights, receiver, bolt (Fig.15), trigger mechanism, magazine, stock and bayonet. Their functions are the same as those of the Type 53 Carbine. The differences are:

1. The backsights are a flip up ladder type. The backsight leaf has graduations ranging from 4 – 24, which indicate 400m to 2400m in range. At ranges under 300m, do not raise the backsight leaf. Take aim using the battle sight notch at the rear of the sight. At 400m range, raise the backsight leaf, slide up the backsight slide and take aim using the sight notch at the base of the backsight leaf. At ranges of 500m and up, raise the backsight leaf and slide the backsight slide up to the desired range indicated on the backsight leaf. Take aim using the sight notch on the backsight slide.

2. The bolt stop is on the left hand side of the receiver.

3. The bolt (Fig.15) has a cover. The safety mechanism is operated by a knob at the rear of the bolt.

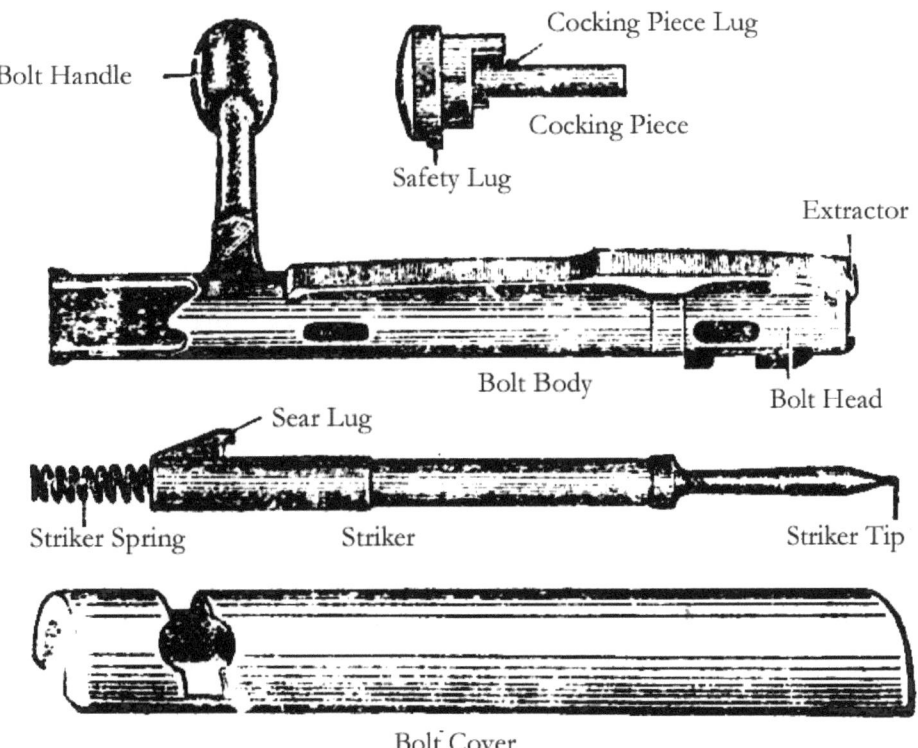

Fig.15  Bolt Assembly

## B. Disassembly and Assembly

### 1. General Notes
As for the Type 53 Carbine.

### 2. Disassembly
(1) **Remove the Cleaning Rod.** Depress the cleaning rod release catch with the thumb of the left hand and lift out the cleaning rod.

(2) **Remove the Bolt.** Hold the rifle face up and pull the bolt stop with the thumb of the left hand. With the right hand, lift up the bolt handle, pull the bolt to the rear and remove the bolt.

(3) **Remove the Magazine Floorplate.** Hold the magazine floorplate lightly with the left hand. With the right hand push forward the floorplate catch and remove the floorplate.

(4) **Disassemble the Bolt.** Hold the bolt body with the left hand with the thumb on the bolt head. With the palm of the right hand push down on the safety knob and rotate it 90 degrees to the right (Fig.16). Release the pressure and safety knob will naturally detach. Remove the striker and striker spring.

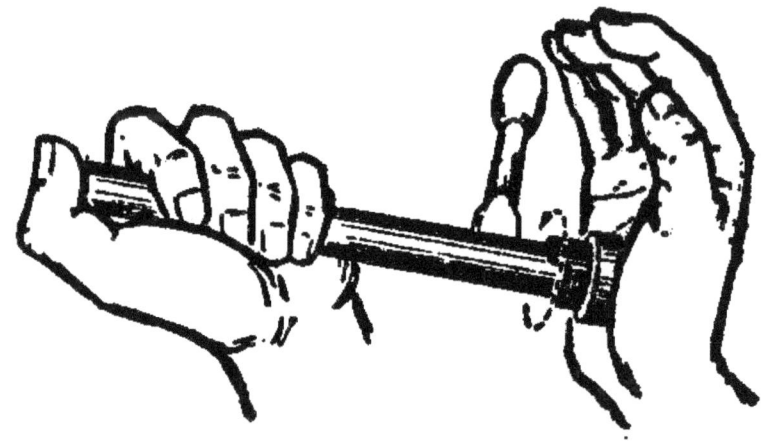

Fig.16 Disassemble the Bolt

### 3. Assembly
Assemble the bolt in the reverse order of disassembly. It should be noted that during bolt assembly: when the striker is inserted into the bolt body, the sear lug must be fitted into rear semi-circular cam groove on the right hand side of the rear of the bolt body. The cocking piece lug on the safety knob must be fitted into the groove on the striker. Press down on the safety knob with the palm of the right hand and rotate it 90 degrees to the left. When fitting the bolt, the bolt cover must first be fitted to the bolt body. The bolt cover must slide into the grooves on both sides of the receiver.

# Type 99 Rifle (Type 99 Arisaka)

The Type 99 Rifle was manufactured in Japan and is an improved version of the 65 Rifle (Fig.17). It is essentially the same as the 65 Rifle, however with the following differences. The barrel is shorter and the calibre is 7.7mm. The backsight is a flip up ladder type with aperture sights. The sling swivel is on the left hand side of the rifle and the rifle has a wire monopod. The words "Type 99" are marked on the top of the receiver.

Fig.17 Type 99 Rifle

# 30 Rifle (M1903 Springfield)

The M1903 Rifle is a modified American version of the German Mauser rifle and was first manufactured in 1903. The calibre is 0.30in (7.62mm) and it is designated the 30 Rifle (Fig.18).

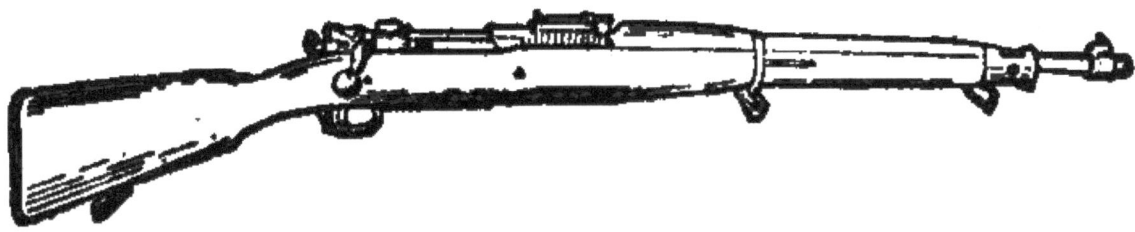

Fig.18 30 Rifle

**A. Description of Parts and Function**
The components of the rifle include the barrel, sights, receiver, bolt (Fig.19), trigger mechanism, magazine, stock and bayonet. Their functions are the same as those of the Type 53 Carbine. The differences are:

1. The backsights are a flip up ladder type. The backsight leaf has graduations ranging from 2 – 27, which indicate 200yd to 2700yd in range (100 yards = 91.4 metres, so simply read the sight indications as multiples of 100 metres). The rear of the backsight base has graduations for adjusting windage, which is adjusted by the windage knob at the front of the backsight base.

2. The bolt stop is on the left hand side of the receiver.

3. The safety catch is on the top of the bolt sleeve and it is used to apply the safety mechanism.

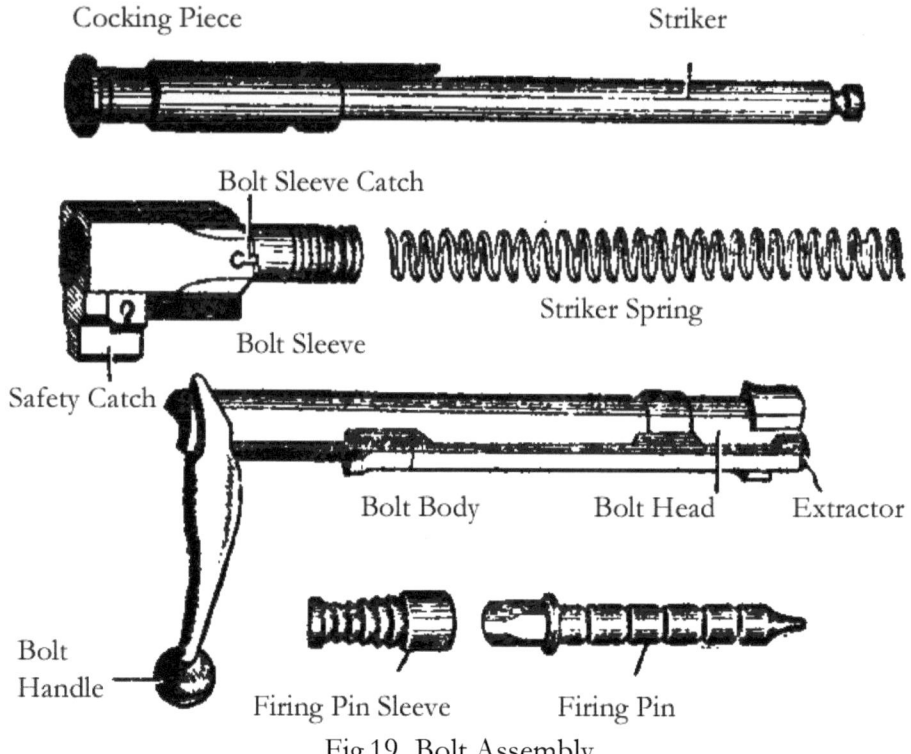

Fig.19 Bolt Assembly

## B. Disassembly and Assembly

### 1. General Notes
As for the Type 53 Carbine.

### 2. Disassembly

(1) **Remove the Bolt.** Rotate the cut off to the middle position. With the right hand, lift the bolt handle upwards and close it down again. Lift the safety catch to the vertical position. With the right hand lift up the bolt handle and pull it to the rear and remove the bolt.

(2) **Remove the Magazine Floorplate.** Insert the end of a cleaning rod into the circular hole in the floorplate and depress the spring loaded floorplate catch. Push the floorplate to the rear to release it from the trigger guard.

(3) **Disassemble the Bolt.** Hold the bolt body with the left hand. Hold the bolt sleeve with the right hand and depress the bolt sleeve catch. Unscrew the bolt sleeve anticlockwise and separate it from the bolt body (Fig.20).

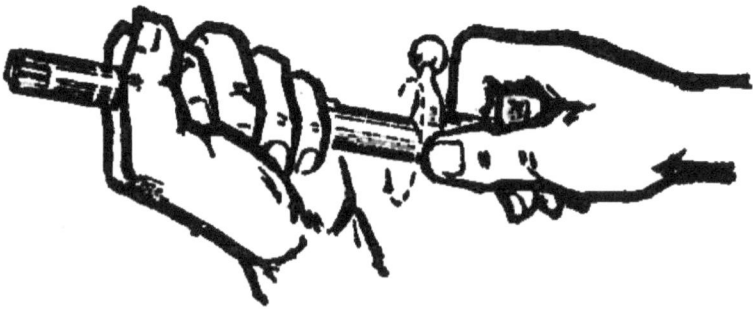

Fig.20 Disassemble the Bolt

Hold the bolt sleeve with the left hand, and with the right hand rotate the safety catch to the left with the thumb and index finger. Hold the striker assembly vertically with the cocking piece on a wooden surface, holding the firing pin sleeve connecting the firing pin and striker between the thumb and index finger of the left hand. Push down with force to compress the main spring. With the right hand, remove the firing pin (Fig.21), by pushing it to the side. Slowly release the pressure on the striker spring, and disassemble the striker, striker spring and bolt sleeve.

### 3. Assembly

Assemble the bolt in the reverse order of disassembly. It should be noted that during bolt assembly: after the bolt sleeve, striker and firing pin have been assembled, the safety catch should be in the vertical position. When the bolt is fitted into the receiver and closed, the safety catch should be turned to the left.

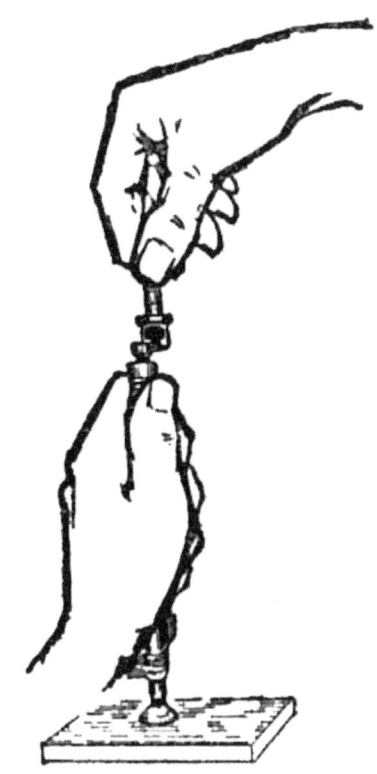

Fig.21 Disassemble the Bolt Sleeve

## 3. Weapons Care and Stoppages / Malfunctions

### A. Weapons Care

1. Rifles and ammunition should be stored separately in a safe, dry and ventilated place. When rifles are not in use, the trigger should be pressed in order to relieve striker spring tension, the backsight slides should be set at the lowest setting, and flip up ladder sights should be folded down. The Type 53 Carbine should have the bayonet folded.

2. Rifles should be cleaned after firing, in the lulls in combat and after duty or training. After firing, clean the rifle in a timely manner and then once a day for three or four days afterwards. When cleaning the rifle, pay particular attention to the rifle bore and operating mechanism, after which the metal parts should be oiled. The rifle should also be oiled during long periods of firing. If the rifle is used after a period of storage, each part should be lightly oiled.

When rifles are brought indoors from the cold, any moisture should be wiped away and the rifle oiled. After immersion in sea water, rifles should first be washed with cold water and then cleaned. After contamination with poison gas or radioactive materials, rifles should be decontaminated in a timely manner. After cleaning and oiling, rifles should be aired in a dry room. Drying by fire or by exposure to the sun is prohibited, in order to protect the wooden stock from warping.

3. Rifles and ammunition should be inspected periodically. It is important to check that the metal parts are free of rust, the sights are undamaged, the bore is clean and the operating mechanism functions normally. Ammunition and rifle accessories should be complete and accountable.

## B. Stoppages / Malfunctions

If stoppages or malfunctions arise during the firing of a rifle, the stoppage or malfunction can be corrected by the following table.

**Table: Stoppages / Malfunctions, Causes and Corrective Actions**

| Stoppage or Malfunction | Probable Cause | Corrective Action |
|---|---|---|
| Failure to fire | 1. Faulty ammunition | 1. Re-cock the rifle or reload a new round |
| | 2. Broken striker or weakened striker spring | 2. Repair or replace striker (or striker spring) |
| | 3. Excessive oil or grease in the operating mechanism | 3. Clean rifle |
| Failure to extract / eject | 1. Dirty chamber or ammunition | 1. Clean chamber or ammunition |
| | 2. Damaged extractor | 2. Replace extractor |
| | 3. Split cartridge case | 3. Remove split cartridge case |
| | 4. Did not open bolt after firing | 4. Open bolt after firing |
| Failure to feed or chamber a round | 1. Dented or deformed cartridge | 1. Reload a new round |
| | 2. Spent case in chamber | 2. Extract case, clean chamber |
| | 3. Weakened magazine spring | 4. Replace magazine spring |
| Safety mechanism fails to engage | 1. Rust on safety mechanism | 1. Oil safety mechanism |
| | 2. Unserviceable safety mechanism | 2. Replace safety mechanism parts |
| | 3. Wear on safety catch, safety lug or lug recess | 3. Replace parts where possible |
| Trigger failure | 1. Wear on sear or sear lug | 1. Replace parts |
| | 2. Weakened trigger spring | 2. Replace trigger spring |

# Submachine Guns

## 1. Combat Capabilities

The submachine gun is the automatic weapon of the Militia for close quarter battle. Submachine guns are effective at killing the enemy at ranges under 200m. They are simple and sturdy in design, lightweight, easy to carry and simple to use. Submachine guns are mainly fired in short bursts (2 - 3 rounds), but they can also be fired in long bursts and in single shots.

## 2. Description of Parts and Function, Disassembly and Assembly

### Type 50 Submachine Gun (PPSh-41)

The 7.62mm calibre, Type 50 Submachine Gun was first manufactured in China in 1950 (Fig.22).

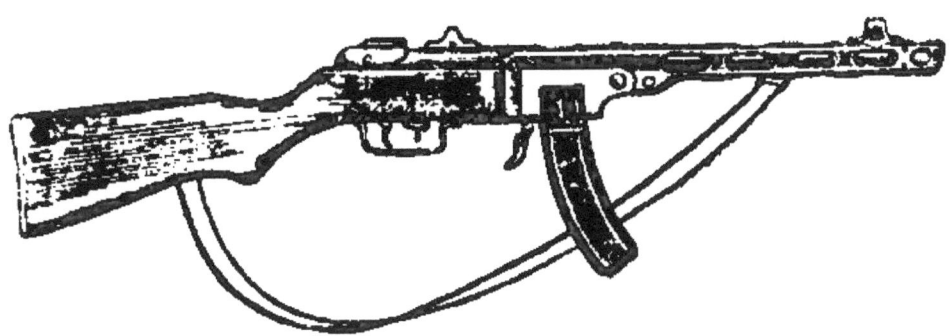

Fig.22 Type 50 Submachine Gun

**A. Description of Parts and Function**

The components of the submachine gun include the barrel, upper receiver, lower receiver, sights, bolt assembly, recoil spring assembly, trigger mechanism, stock, magazine and accessories kit.

1. **Barrel.** The barrel directs the flight of the bullet. The interior of the barrel consists of the rifled bore and the chamber. The chamber houses the cartridge and the rifling imparts spin on the bullet to provide accuracy and penetration. The rifling of the bore has four grooves and the distance between two opposite lands is the calibre of the submachine gun.

2. **Upper Receiver.** (Fig.23) The upper receiver houses the barrel and trunnion. The upper receiver consists of the barrel hand guard, muzzle compensator and receiver catch. The barrel hand guard protects the barrel and prevents the burning of the hand during firing; the muzzle compensator reduces muzzle climb when firing and the receiver catch separates the upper receiver and the lower receiver.

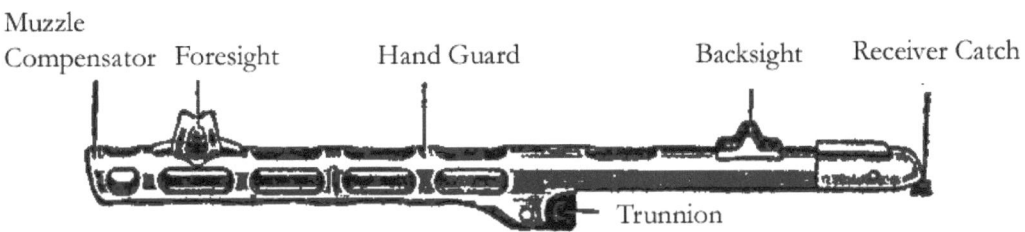

Fig.23 Upper Receiver

15

3. **Sights.** The sights are for aiming the submachine gun. The sights consist of the backsight and foresight. The backsight has two aperture sights. The low sight setting (sight flipped forward) is for ranges up to 100m. The high sight setting (sight flipped rearward) is for ranges from 100m to 200m.

4. **Lower Receiver.** (Fig.24) The lower receiver houses the component parts of the operating mechanism. The lower receiver consists of the ejector, magazine housing and the magazine catch. The ejector serves to eject spent cartridge cases and the magazine housing and magazine catch secures the magazine.

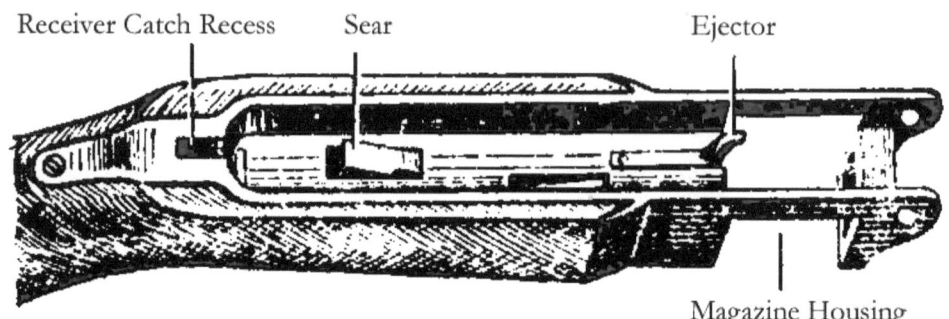

Fig.24 Lower Receiver

5. **Bolt Assembly.** (Fig.25) The bolt assembly is used for feeding the cartridge into the chamber, closing the breech, firing the shot and extracting the cartridge case. The bolt assembly consists of the bolt, firing pin, extractor and cocking handle. The bolt's features are: the feed rib at the front of the bolt, which feed cartridges from the magazine into the chamber; the sear notch at the rear of the bolt, which through the control of the sear, regulates the firing cycle; the firing pin which strikes the cartridge primer; and the extractor which extracts spent cases from the chamber. The cocking handle is pulled back to cock the submachine gun and the safety catch is located at the top of the cocking handle. It can be used to lock the bolt in a forward or rearward position.

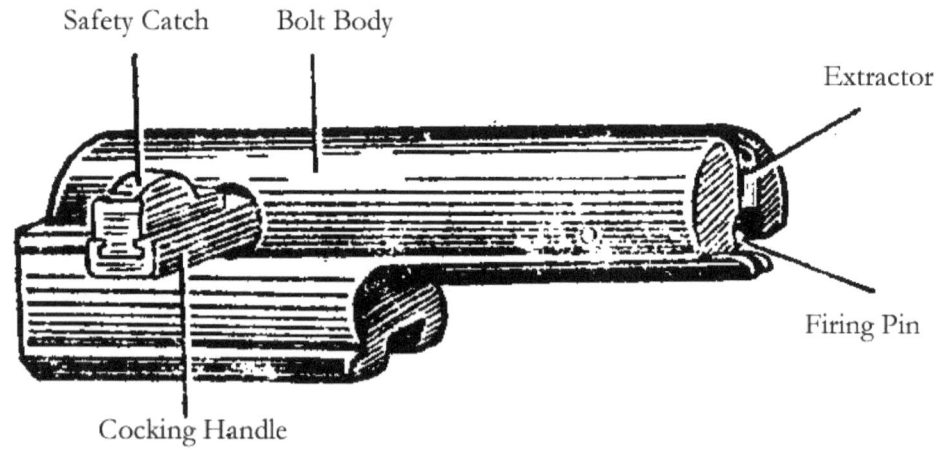

Fig.25 Bolt Assembly

6. **Recoil Spring Assembly.** (Fig.26) The recoil spring assembly returns the bolt to the forward position during the firing cycle. The recoil spring assembly consists of the recoil spring, guide rod and buffer. The recoil spring serves to provide tensile force to return the bolt to the forward position. The guide rod prevents the recoil spring from bending during recoil and the buffer serves to protect the rear of the bolt from impact when it travels to the rear of the receiver during the firing cycle.

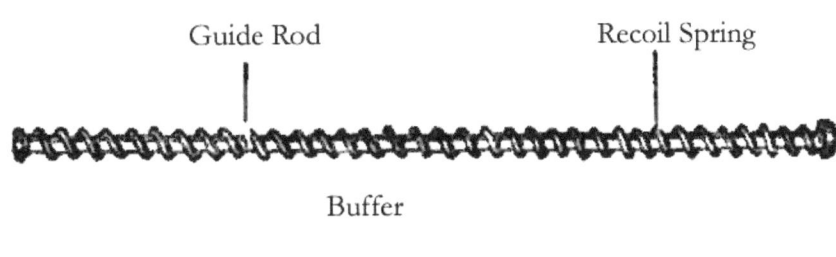

Fig.26 Recoil Spring Assembly

7. **Trigger Mechanism.** (Fig.27) The trigger mechanism is used for firing single shots and bursts. The trigger mechanism consists of the trigger guard, selector, sear and trigger. The trigger guard houses the trigger mechanism. The selector is used to select single shot or automatic fire. Single shot is selected by pulling the selector to the rear and automatic is selected by pushing the selector forward. The sear controls the firing cycle and the trigger is used to fire the submachine gun.

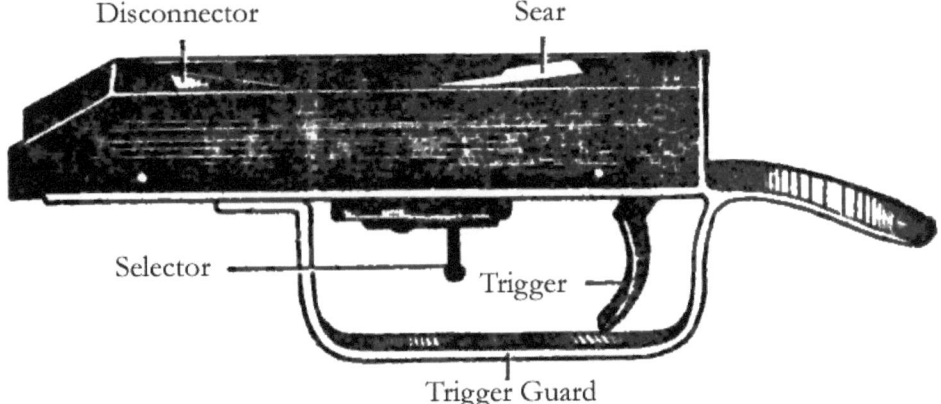

Fig.27 Trigger Mechanism

8. **Stock.** The stock is for carrying and operating the submachine gun. The accessories are stored within the butt.

9. **Magazine and Accessories Kit.** The magazine holds the cartridges. It consists of the magazine body, magazine platform, spring and floorplate. The submachine gun's accessories kit consists of a cleaning rod, screwdriver tool, punch, bristle brush, oil bottle, sling and magazine pouches.

## B. Disassembly and Assembly

### 1. General Notes
As for the Type 53 carbine.

### 2. Disassembly
(1) **Remove the Magazine.** Hold the receiver under the backsights with the left hand. Hold the magazine with the right hand and press the magazine catch with the thumb, and then pull the magazine down.

(2) **Separate the Upper and Lower Receivers.** Hold the hand grip of the butt stock with the right hand and press the receiver catch forward with the thumb. Simultaneously swing the barrel down with the left hand (Fig.28).

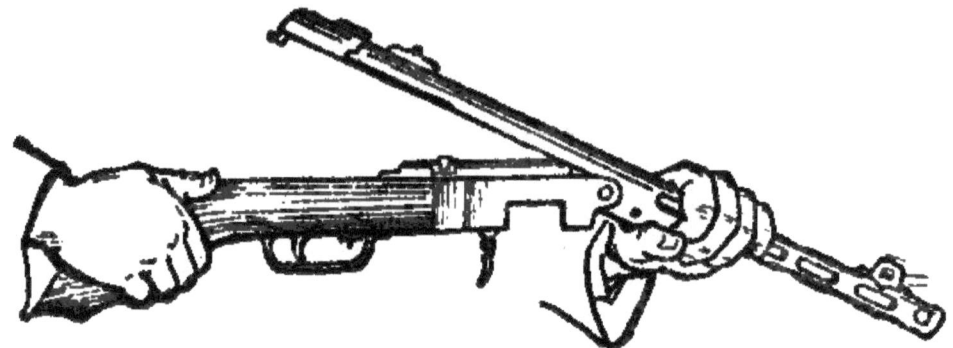

Fig.28 Separate the Upper and Lower Receivers

(3) **Remove the Bolt and Recoil Spring Assemblies.** Hold the front of the stock with the left hand. Pull the bolt to the rear with the right hand and simultaneously lift the front of the bolt upwards and to the right. Remove the bolt and recoil spring assemblies (Fig.29).

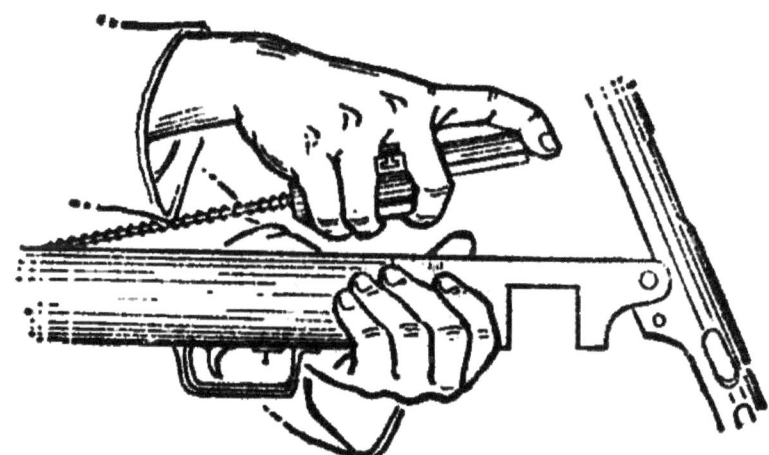

Fig.29 Remove the Bolt and Recoil Spring Assemblies

## 3. Assembly
Assemble the submachine gun in the reverse order of disassembly.

## Type 54 Submachine Gun (PPS-43)

The 7.62mm calibre, Type 54 Submachine Gun was first manufactured in China in 1954 (Fig.30).

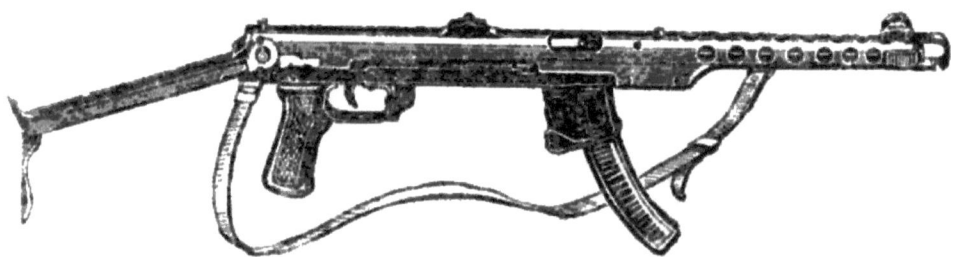

Fig.30 Type 54 Submachine Gun

## A. Description of Parts and Function

The components of the submachine gun include the barrel, upper receiver (Fig.31), lower receiver, sights, bolt assembly (Fig.32), recoil spring assembly (Fig.33), trigger mechanism, stock, magazine and accessories kit. Their functions are the same as those of the Type 50 Submachine Gun. The differences are:

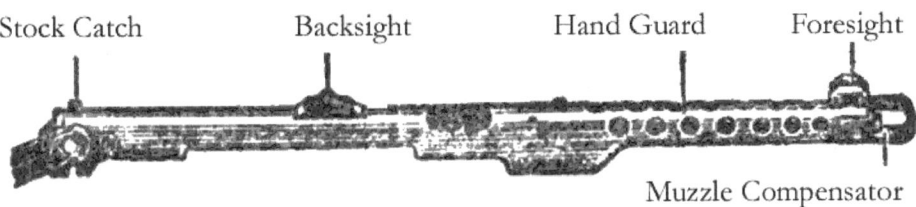

Fig.31  Upper Receiver

Fig.32  Bolt Assembly

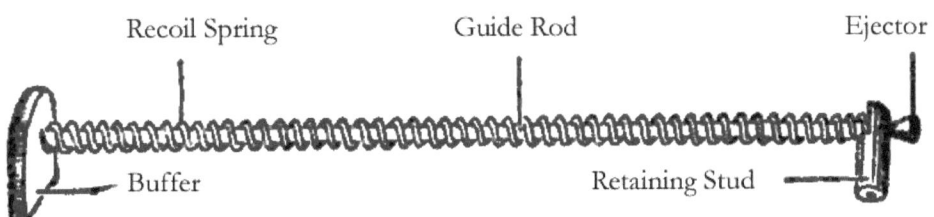

Fig.33  Recoil Spring Assembly

1. The stock release catch is located at the rear of the upper receiver.

2. The foresight is an adjustable type, which can be raised or lowered to correct for elevation, or moved left or right to correct for windage.

3. The cocking handle does not have a safety catch.

4. The lower receiver and trigger mechanism (Fig.34). The lower receiver connects with the upper receiver and the houses the trigger mechanism. The lower receiver consists of the safety catch, receiver catch, magazine housing, magazine catch and pistol grip. The safety catch is on the right of the trigger guard. Pull the safety catch to the rear to engage the safety mechanism.

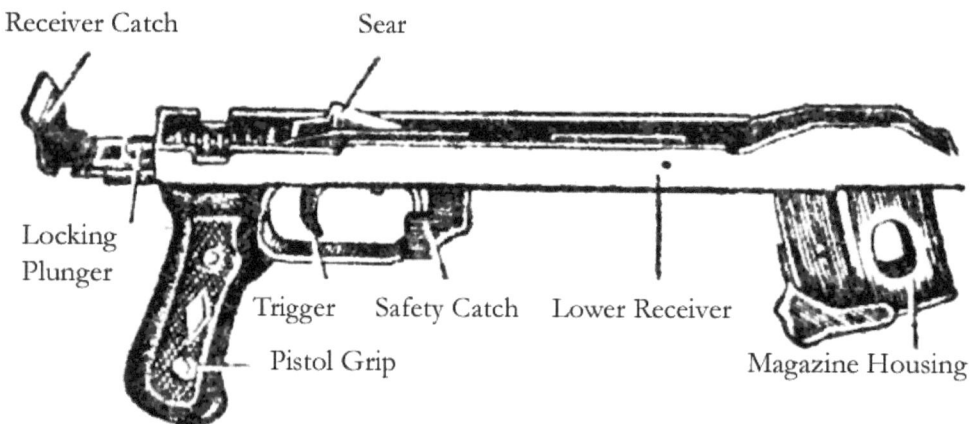

Fig.34 Lower Receiver and Trigger Mechanism

5. The stock is a folding type and consists of the steel stock and the butt plate.

6. The accessories kit includes a cleaning rod with a handle featuring an integral foresight key and punch.

**B. Disassembly and Assembly**

**1. General Notes**
As for the Type 53 Carbine.

**2. Disassembly**
(1) **Unfold the Stock.** Hold the receiver under the backsights with the left hand. Press the stock catch with the thumb of the right hand and push open the stock with the fingers of the right hand.

(2) **Remove the Magazine.** Hold the receiver under the backsights with the left hand. Hold the magazine with the right hand and press the magazine catch with the thumb, and then pull the magazine down.

(3) **Separate the Upper and Lower Receivers.** Hold the barrel hand guard with the left hand. Hold the pistol grip with the right hand and press the receiver catch forward, and swing the barrel down (Fig.35).

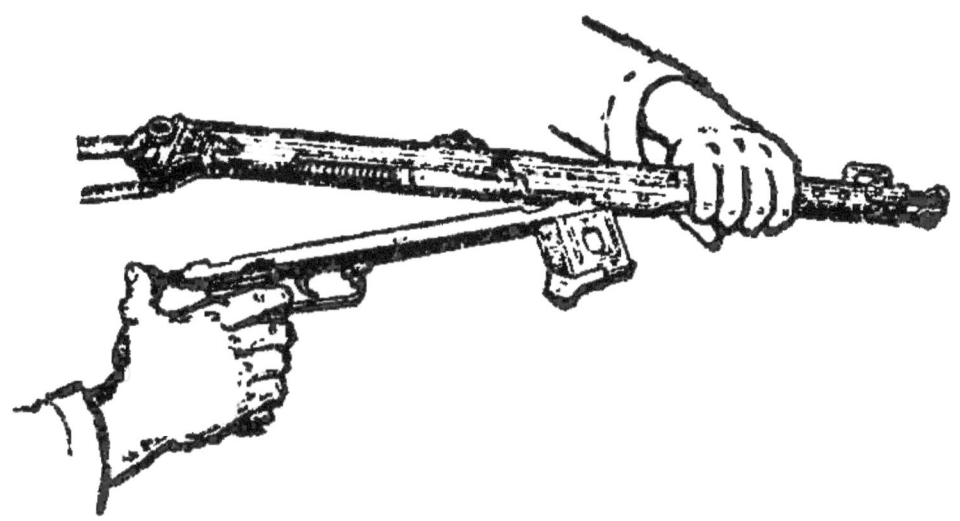

Fig.35 Separate the Upper and Lower Receivers

(4) **Remove the Bolt and Recoil Spring Assemblies.** Hold the barrel hand guard and magazine housing with the left hand. Pull the bolt slightly to the rear with the right hand and then pull the bolt and recoil spring assemblies downwards and to the right to remove them from the receiver (Fig.36).

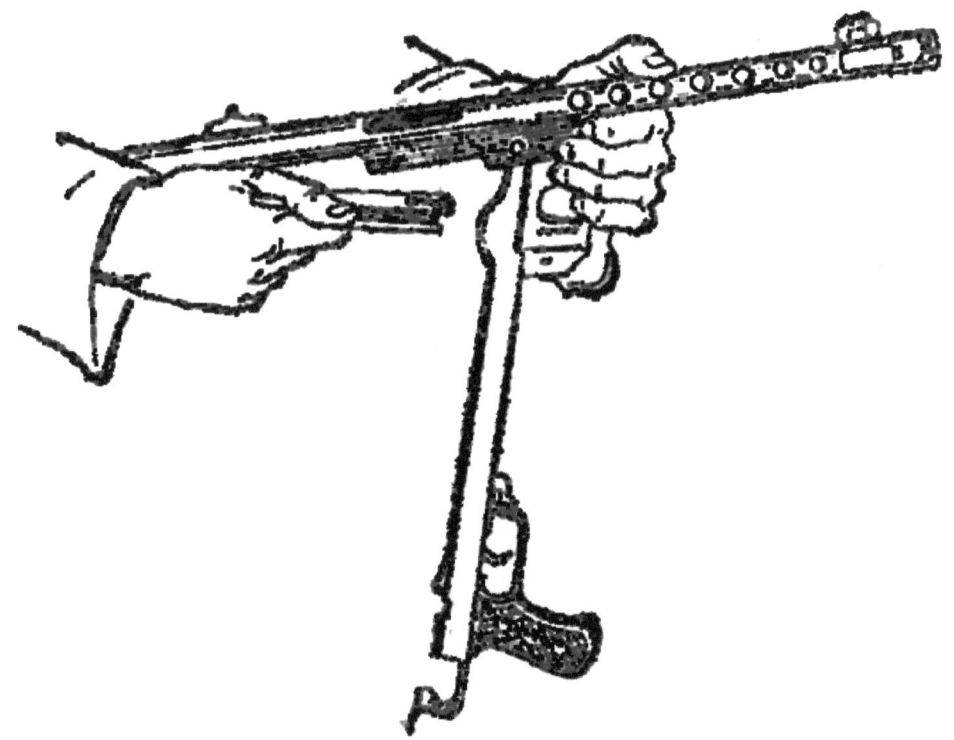

Fig.36 Remove the Bolt and Recoil Spring Assemblies

## 3. Assembly
Assemble the submachine gun in the reverse order of disassembly.

# 3. Weapons Care and Stoppages / Malfunctions

## A. Weapons Care
As with rifles.

## B. Stoppages / Malfunctions
If stoppages or malfunctions arise during the firing of a submachine gun, the stoppage or malfunction can be corrected by the following table.

### Table: Stoppages / Malfunctions, Causes and Corrective Actions

| Stoppage or Malfunction | Probable Cause | Corrective Action |
|---|---|---|
| Failure to feed or chamber a round | 1. Damaged magazine or weakened magazine spring<br><br>2. Dirty chamber and bolt | 1. Replace magazine<br><br>2. Clean chamber and bolt |
| Failure to fire | 1. Faulty ammunition<br><br>2. Worn firing pin tip<br><br>3. Weakened recoil spring | 1. Reload a new round<br><br>2. Repair or replace firing pin<br><br>3. Replace recoil spring |
| Failure to extract / eject | 1. Extractor damaged or extractor spring weakened<br><br>2. Bolt, upper receiver, chamber or ammunition dirty | Remove the magazine, pull the cocking handle to the rear to extract and eject the cartridge case<br><br>1. Replace extractor or extractor spring<br><br>2. Clean receiver, bolt, chamber or ammunition |
| Safety mechanism fails to engage | 1. Rust on safety mechanism<br><br>2. Unserviceable safety mechanism<br><br>3. Wear on safety catch, safety lug or lug recess | 1. Oil safety mechanism<br><br>2. Replace safety mechanism parts<br><br>3. Replace parts where possible |
| Trigger failure | 1. Wear on sear notch or sear tip<br><br>2. Weakened trigger spring | 1. Replace parts<br><br>2. Replace trigger spring |

# The Theory of Small Arms Fire

## 1. Small Arms Fire and Recoil

### A. Small Arms Fire

Small arms fire is the firing of a bullet from a gun barrel, by the pressure of gases released by the detonation of the propellant in the cartridge. In order to understand the process of small arms fire, it is first important to understand the design of a cartridge.

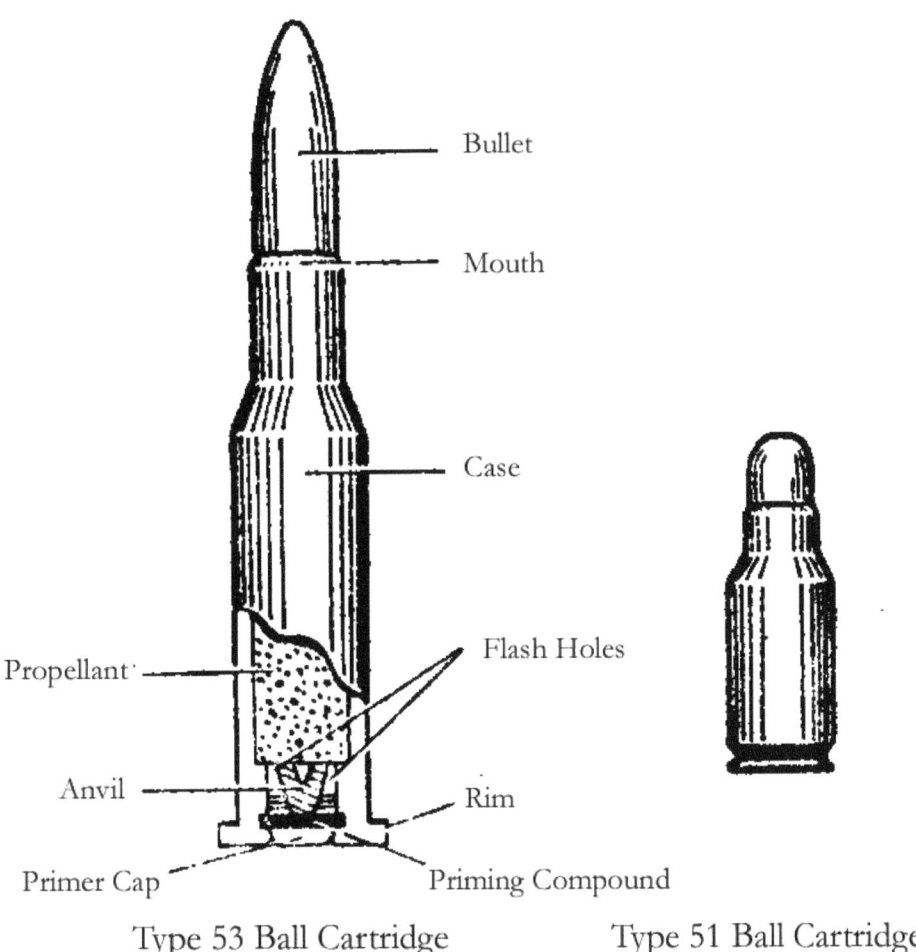

Fig.37  Cartridge Design[7]

**Cartridge Design.** A cartridge consists of a bullet, case, primer and propellant (Fig.37). The propellant is contained within the case and the primer is located at the centre of the base of the cartridge. The primer contains the priming compound and there are two flash holes that connect to the interior of the case.

**Ignition Process.** When a weapon is fired, the striker tip or firing pin strikes the primer of the cartridge and the priming compound is ignited. This ignition flashes through the flash holes in the base of the cartridge and through to the case. This in turn ignites the propellant in the case and creates a large volume of gas at high pressure. This gas forces the bullet to separate from the case and accelerate down the barrel rifling, which imparts rotation on the bullet, until it is pushed out of the muzzle.

---

[7] Translator's Note: The Type 53 Ball and Type 51 Ball cartridges were the Chinese designations for the 7.62x54mmR ball and 7.62x25mm ball cartridges respectively.

## B. Recoil

Recoil is the rearward movement of a weapon when it is fired.

**Creation of Recoil.** When a weapon is fired, the propellant gases exert pressure in all directions (Fig.38). The pressure is contained by the walls of the chamber and barrel, and the base of the cartridge and the bolt, so that the bullet is propelled forward by the gas pressure. The recoil is the rearward moving force imparted on the weapon by the forward movement of the bullet and gases.

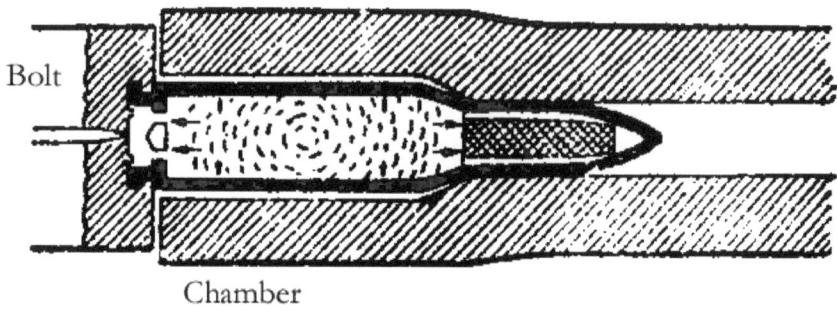

Fig.38 Function of Propellant Gases

**Effect of Recoil on Accuracy.** The weapon's recoil begins simultaneously with the forward movement of the bullet. As the bullet accelerates, the amount of recoil also gradually increases. However, due to the short duration of time that the bullet is in the barrel, and the several hundred fold difference in relative weights between the weapon and the bullet, there is only about 1mm of movement of the weapon due to recoil before the bullet leaves the barrel. This recoil is essentially a rearwards movement, so it has little effect on accuracy. However at the moment that the bullet leaves the muzzle, the reactive force that is created by the forward expulsion of propellant gases from the muzzle results in a distinct increase in recoil. However by this time, the bullet has already left the muzzle, so there is little effect on the accuracy of a single shot. During automatic fire however, the movement due to recoil changes the original point of aim, so that after the first shot is fired, the accuracy of subsequent shots is affected. However, if the firer simply corrects the aim and adapts to the movement of the weapon during automatic fire, then the effect of recoil can be reduced, and the accuracy of fire is improved.

# 2. Practical Implications of Trajectory

## A. Trajectory

Trajectory is the path of a bullet in flight. A bullet does not travel in a straight line after it leaves the muzzle. During its flight it is affected by the actions of gravity and air resistance, and it gradually falls and slows as it travels. The bullet forms an uneven arc with a relatively long and straight rising arc and a short and curved falling arc (Fig.39).

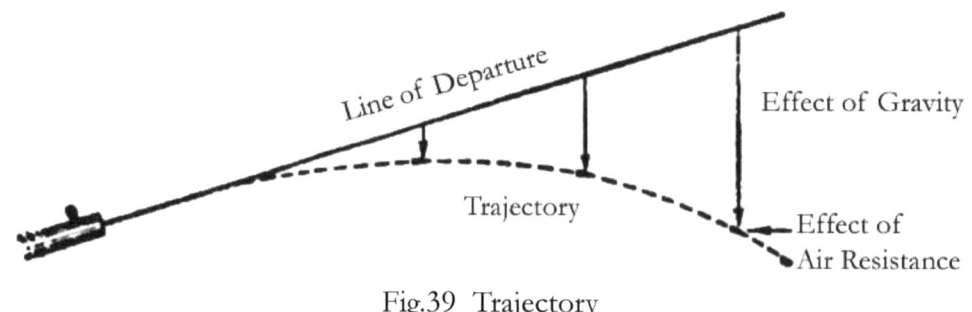

Fig.39 Trajectory

## B. Dangerous Space

The dangerous space is the space in the line of fire where the height of the trajectory does not exceed the height of the target.[8] The size of the dangerous space depends on the height of the target, the 'flatness' of the trajectory and the position of the target in the terrain. The taller the height of the target, the shorter the range of the target, or the flatter the terrain; then the larger the size of the dangerous space and the easier it is hit the target. The higher the target is sited relative to the weapon, the longer the range of the target, or the more uneven the terrain (eg depressions); then the smaller the size of the dangerous space and the harder it is to hit the target (Fig.40).

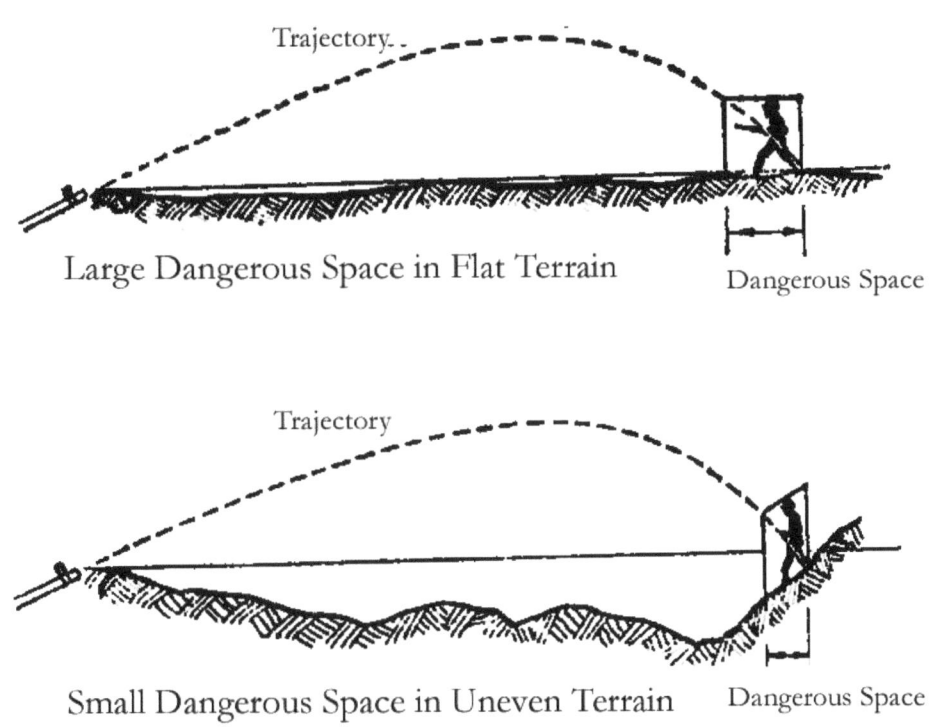

Fig.40 Relationship between Terrain and Dangerous Space

## C. Defilade and Dead Ground

Defilade is the space between the apex of an obstacle in the direct line of fire through which the bullet cannot pass, and the point of impact of the bullet. There are two zones within a defilade; these are the dead ground and the dangerous space. A target which is in defilade and is unable to be hit by direct fire is said to be in the dead ground. A target which is in defilade and is able to be hit by direct fire is said to be in the dangerous space (Fig.41). The taller the height of the obstacle creating the defilade, or the lower the target is sited relative to the weapon, then the greater the size of the dead ground. Conversely, the shorter the height of the obstacle, or the higher the target is sited relative to the weapon, then the smaller the size of the dead ground.

---

[8] Translator's note: In other words, the dangerous space is the distance between the point where the bullet would first strike the target ('first catch') and the point where it would first strike the ground ('first graze'). In this space, the height of the trajectory cannot be higher than the target, in order to for the target to be hit.

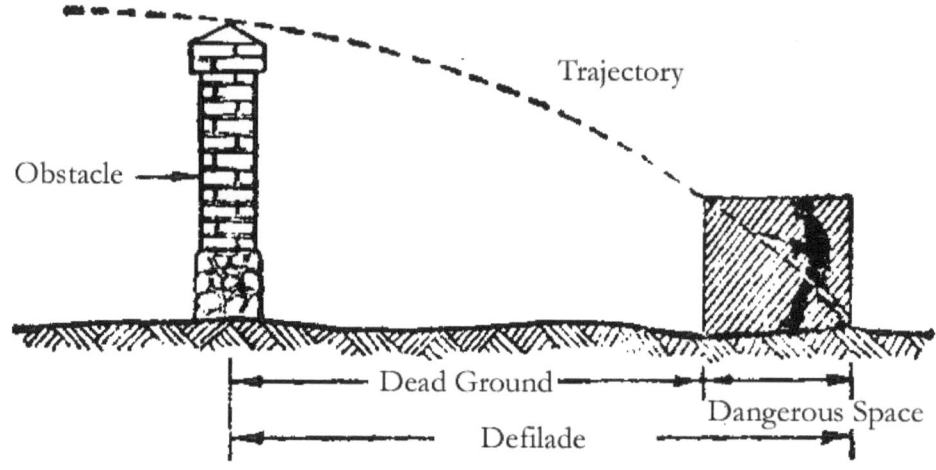

Fig.41  Defilade and Dead Ground

Understanding the meaning and significance of dangerous space, defilade and dead ground is important in battle.  It will allow you to better use the terrain to conceal your approach on the enemy and reduce the effectiveness of enemy fire.  It will allow you to select an appropriate firing position to destroy the enemy with direct fire, or indirect fire if the enemy is in dead ground.

# 3. Sighting and Aiming

## A. Using the Sights

As a bullet flies through the air, it is affected by the actions of gravity and air resistance, and it gradually falls and slows as it travels.  If the target is fired on by sighting through the bore, the bullet will fall short of the target (Fig.42).

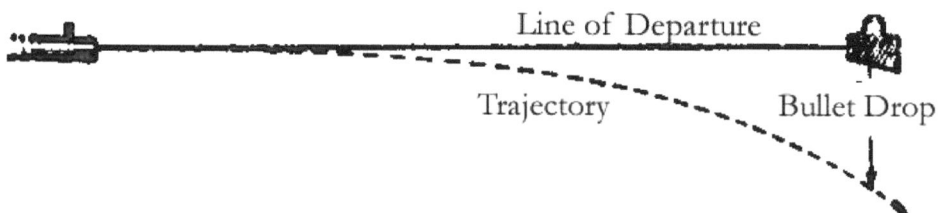

Fig.42  Trajectory of a Shot Sighted Through the Bore

In order to hit the target, the muzzle must be elevated, and the weapon and line of sight held at a specific angle.  At any given range, a target can be hit by aiming with the range indication on the backsight slide set to correspond with the range to the target (Fig.43).  Therefore, the purpose of the sights is to raise the weapon to a specific angle of elevation when firing at targets at a specific range.

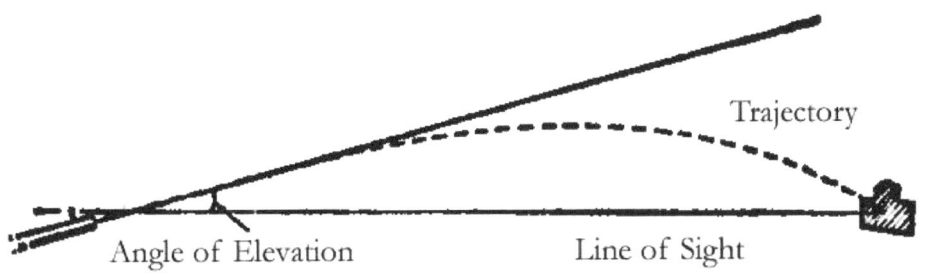

Fig.43 Trajectory of a Shot Sighted Through the Weapon Sights

## B. Sight Radius, Line of Sight, Point of Aim and Height of Trajectory
**Sight Radius.** The distance along the central sighting line between the backsight and tip of the foresight.
**Line of Sight.** The central sighting line through the backsight, the tip of the foresight and extending to the target.
**Point of Aim.** The point at the end of the line of sight which is being targeted.
**Height of Trajectory.** The vertical distance between the line of sight and any point on the trajectory.

## C. Setting the Sight and Selecting the Point of Aim
In order for the bullet to hit the target accurately, the firer should select the appropriate backsight range setting and point of aim, based on the target's range, size and height of trajectory. (See Appendix 3 & 4)

Procedure:
1. If the target's range is a multiple of 100m, set the sight to the appropriate range. The point of aim is the centre of the target.

2. If the target's range is not a multiple of 100m, set the sight to the nearest hundred metres above or below the actual range. Lower or raise the point of aim respectively.

3. If the target is within 300m range, generally the sight can be used set at '3'. When using this sight setting, the point of aim is: below or under, for a small target; and in the centre, for a large target.

## D. Observing the Shot and Correcting the Point of Aim

Errors in determining the range and the effect of external factors may result in the shot missing the target. Therefore the firer should observe the fall of shot and determine whether the target has been hit or missed. The determination can be made by the firer observing for the strike of the bullet on the ground or in the water, the trail of a tracer round, or a change in the state of the target. Misses should be carefully analysed to identify the cause of the miss and quickly corrected. The amount of correction of the aim will be determined by the degree of the miss. To correct a horizontal point of impact, adjust the point of aim by aiming off (when correcting the point of impact by adjusting a windage adjustable backsight, the point of aim is unchanged). If correcting a vertical point of impact, the sight setting can be adjusted to correct the range.

# 4. The Effect of External Factors on the Shot and the Point of Aim

## A. The Effect of Wind on the Shot

### 1. Determining Wind Direction and Strength

The direction of the wind can be classified according to its angle, relative to the direction of the shot. These are: cross wind, oblique wind, tail wind and head wind. Wind can be classified according to strength:

**Fresh to Strong Breeze.** A wind speed of 8 – 12 m/s (Beaufort 5 – 6). Indicators: flags will be flying horizontal relative to their poles and will make a strong flapping sound; grass will be bent towards the ground and large tree branches will shake.

**Gentle to Moderate Breeze.** A wind speed of 4 -7 m/s (Beaufort 3 -4). Indicators: flags will be fluttering open and will make a clear fluttering sound; grass will wave continuously and small tree branches will sway.

**Light Breeze.** A wind speed of 2 -3 m/s (Beaufort 2). Indicators: flags will flutter lightly; grass and small tree branches will move slightly.

### 2. Allowing for Wind and Correcting the Point of Aim

Cross winds will cause the bullet to deviate from its natural trajectory. The greater the wind strength, the greater the deviation will be and the greater the distance the shot will miss the target (Fig.44).

In order to hit the target accurately, the point of aim or the windage adjustable backsight must be adjusted into the direction of the wind. The table below shows the amount of correction for a gentle to moderate breeze. For a fresh to strong breeze, double the amount of correction. For a light breeze or oblique wind, halve the amount of correction. The amount of correction is determined from the centre of the target. The point of aim is not changed after a windage adjustable backsight has been adjusted.

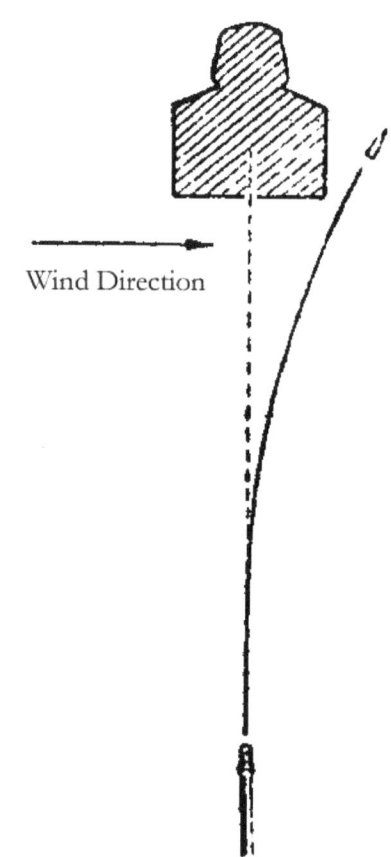

Wind Direction

Fig.44 Effect of Cross Wind on Trajectory

**Table: Correction for a Moderate Cross Wind**

| Range (m) | Amount of Correction of Point of Aim | |
|---|---|---|
| | cm | Body Widths |
| 100 | 3 | - |
| 200 | 9 | - |
| 300 | 20 | 0.5 |
| 400 | 40 | 1.0 |
| 500 | 68 | 1.5 |

Tail winds will cause the trajectory to deviate high and long. Head winds will cause the trajectory to deviate low and short. If the wind speed is less than 10m/s, the effect will be minimal and correction is not required if the range is within 400m. If the range is longer than 400m, then lower or raise the point of aim as required.

## B. The Effect of Sunlight on the Point of Aim

When taking aim in direct sunlight, the backsight notch will cast shadows and create the optical illusion of three backsight notches in the sight picture. If the aim is taken with the illusionary 'ghost' sight notch, the trajectory will deviate towards the direction of the sunlight. If the illusionary dark 'shadow' sight notch is used, the trajectory will deviate in the opposite direction of the sunlight (Fig.45). Therefore, practise taking aim in different levels of sunlight. Practise taking aim in and out of the shade, in different types of shade and light, and learn to determine the true sight picture.

When taking aim, do not hold the sight picture for too long, in order to avoid blurring of the eyes and inducing errors in the aim. Protect the sights and do not polish them or allow them to be reflective.

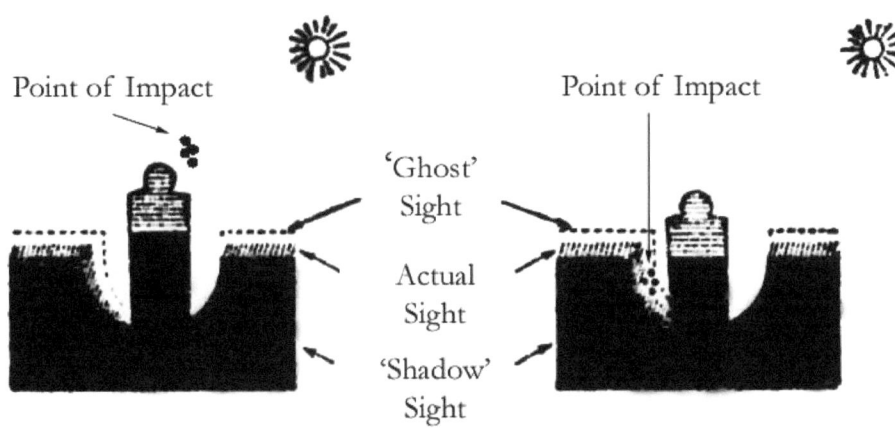

Fig.45  Effect of Sunlight on the Point of Aim

# Determining Range

## 1. Determining Range by Pace Counting

Face the target and walk directly to it with normal paces (typically 75cm), counting the number of double paces (a set of two steps, one of each foot). Then take the number of double paces and add half of that number; or multiply the number of double paces by 1.5; to calculate the range in metres.

When pace counting, the size of the paces should be consistent. In uneven terrain, the size of the paces should be increased appropriately. If there are areas of terrain that cannot be measured by pacing, measure the distance by other means and continue pace counting after bypassing the area. Add the two measurements to calculate the total range.

## 2. Determining Range by Eye

### A. Determining Range by Comparing Distances

Compare the distance to the target with distances in a reference terrain deeply impressed in the memory (eg power line poles 50m apart, 100m of the rifle range etc) or a terrain with a known distance, to determine the range. The distance to the target can be divided into equal segments of terrain, which can then be compared with a reference distance terrain in the memory and then multiplied to determine the total range.

It is easy to make the mistake of overestimating the range on a small and distant target, so that it seems farther away than it actually is. When there are unseen or partially obscured depressions (eg valleys, hollows or creeks) in the line of sight, it is easy to make the mistake of underestimating the range, so that it seems closer than it actually is.

### B. Determining Range by Degrees of Visibility

In good weather, range can be determined by differentiating features of targets and terrain, by virtue of their differing degrees of visibility at different distances.

In general terms, at:

| | |
|---|---|
| 200m | - can distinguish individual faces, clothing and equipment; individual strands of barbed wire. |
| 300m - 400m | - can distinguish the colour of clothing; the type of weapons being carried. |
| 500m - 600m | - can clearly distinguish individual human outlines and the movement of arms and legs; the stakes of barbed wire obstacles. |

Errors can arise due to differing conditions such as the weather, the dispositions of targets, the surrounding terrain and backgrounds etc (see Table below). Therefore, it is important to become familiar with these factors and practise determining ranges under differing conditions.

| Range Seems Closer | Range Seems Farther |
|---|---|
| Large targets; lone targets; brighter targets (eg white coloured); clear visibility | Small targets; dull targets (eg grey and dark coloured) |
| Target and background is distinct (eg targets in snow) | Target and background is indistinct (eg dark colours in grass); complex terrain |
| Targets: in clear skies; on water; uphill | Targets in: rain, fog, dusk, dawn, night, downhill |

## C. Determining Range by Parallax

Face the target and extend the right arm about 60cm with the thumb held up. Close the left eye and align the right eye, thumb and target. Do not move the thumb or head. Close the right eye and open the left eye. Estimate the distance in metres between the target and the 'jump point' where the thumb now appears viewed with the left eye. Multiply this distance by 10 to determine the range (Fig.46).[9]

Example: If the distance between the target (pagoda) and the 'jump point' feature (tree) is 25m, then the range is 25m x10 = 250m.

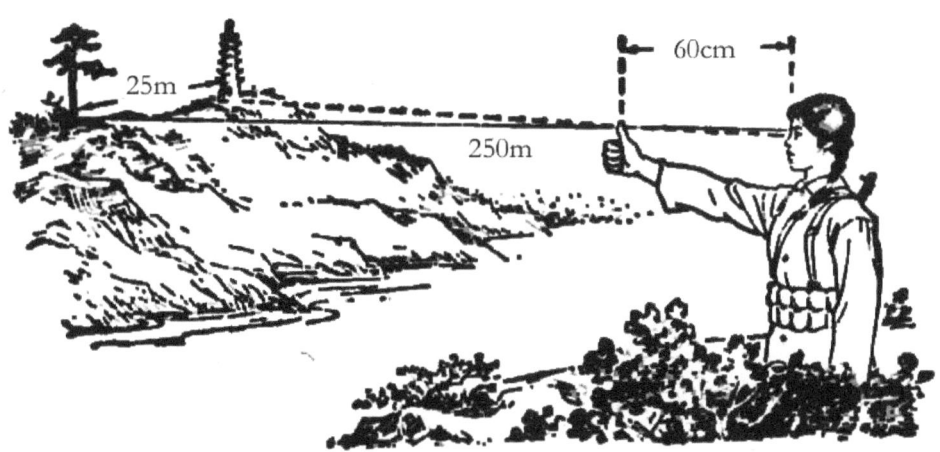

Fig.46 Determining Range by Parallax

---

[9] Translator's Note: This method of determining range is known in Chinese as the 'eye jump' method.

# Marksmanship

# Rifles

## 1. Safety Inspection

For the purposes of safety, rifles must be inspected before and after use, and whenever else it is required. The chamber and magazine must be cleared of ammunition. Drill cartridges must be inspected and confirmed as drill cartridges. During the inspection, pointing the muzzle at another person is strictly prohibited.

On the command "SAFETY INSPECTION", incline the body to the right, pivoting on the right foot and step the left foot one pace forward (one shoulder width). Extend the rifle forward with the right hand and simultaneously grasp the rifle in front of the magazine with the left hand. The upper left arm should be held close to the left ribs and the rifle butt held near the hip. The bayonet should be held slightly higher than eye level (or the muzzle slightly higher than shoulder height). Open the magazine floorplate with the right hand (not with older style rifles)[10], then grasp the bolt handle and look to the front. When the commander inspects the rifle, open the bolt and slightly raise the rifle upwards and to the right, so that the commander can see the chamber. After inspection, in your own time, close the bolt and the magazine floorplate, press the trigger and hold the hand grip with the right hand.

On the command "SAFETY INSPECTION COMPLETE", grasp the rifle at the backsight with the right hand and simultaneously incline the body to the left. Close the right foot to the left foot and recover to the position of attention.

When inspecting drill cartridges, grip the rifle between the legs and remove each round individually for inspection.

## 2. Loading and Sight Setting

### A. Prone Position

On the command "PRONE - LOAD", raise the rifle with the right hand, step the left foot one pace forward, stretch the left arm out in the direction of the left foot with palm down and fingers outstretched slightly to the right. Quickly go down on the left knee and lay on the ground, resting on the left hand side of the body supported by the left elbow. Point the rifle towards the target with the right hand and hold the rifle with the left hand under the backsight and rest the butt on the ground. With the right hand, open the bolt, open an ammunition pouch, remove a charger of cartridges and insert it into the charger slot of the receiver. Place the thumb or index finger of the right hand on the uppermost cartridge in the charger and push the cartridges into the magazine (Fig.47). Close the bolt and return the empty charger clip to the ammunition pouch. Pull the rifle backwards and with the right thumb and index finger, pinch the backsight slide catch and set the slide to the required range (for a rifle with a flip up ladder sight, raise the backsight leaf if required). Move the right hand back to the hand grip, with the rifle face up and the spread both legs so that the feet are shoulder width apart and pointed outwards. The body should be at a 30 degree angle to the target. Watch the front and prepare to fire.

On the command "UNLOAD - STAND", lean the body slightly to the left and lift the bolt handle with the right hand and pull it back with the thumb. Catch the cartridge or case extracted from the chamber with the index and middle fingers. Push down the uppermost cartridge in the magazine down past the interrupter.

---

[10] Translator's Note: All rifles in this manual other than the Type 53 (or other Mosin Nagant types).

Open the magazine floorplate and catch the falling cartridges (on older style rifles, hold the rifle under the magazine with the left hand, and with the right hand work the bolt forward and backward, ejecting cartridges with each cycle). Place the cartridges into an ammunition pouch and close the magazine floorplate. Close the bolt and set the backsight slide to the lowest setting (fold down a flip up ladder sight). Grasp the rifle with the right hand in front of the backsight and draw it backwards. At the same time, roll the body to the right, allow the left leg to naturally bend back, and slightly draw in the left forearm. Using the left hand, knee and leg for support, stand up and simultaneously step one large pace forward with the right foot. Close the left foot to the right foot and recover to the position of attention.

Fig.47 Loading in the Prone Position

## B. Kneeling Position

On the command "KNEEL – LOAD", step forward one pace with the left foot, extend the rifle towards the target with the right hand and hold the rifle under the backsight with the left hand. Kneel down on the right knee and sit on the heel of the right foot. The lower left leg should be nearly vertical and both legs should be held at close to a 90 degree angle. Rest the left forearm on the left thigh. Load the rifle (Fig.48), set the sight and hold the hand grip with the right hand. Watch the front and prepare to fire.

On the command "UNLOAD – STAND", unload the rifle and return the sight to the lowest setting. Hold the rifle in front of the backsight with the right hand, and stand up. Close both feet together and recover to the position of attention.

Fig.48 Loading in the Kneeling Position

## C. Standing Position

On the command "STAND – LOAD", incline the body to the right, pivoting on the right foot and step the left foot one pace forward (one shoulder width). Balance the body's weight equally on both feet and simultaneously extend the rifle forward with the right hand and hold the rifle under the backsight with the left hand. The upper left arm should be held close to the left ribs and the rifle butt held near the hip. The bayonet should be held slightly higher than eye level (or the muzzle slightly higher than shoulder height). Load the rifle, set the sight, and hold the rifle by the hand grip with the right hand. Watch the front and prepare to fire (Fig.49).

On the command "UNLOAD", unload the rifle and return the sight to the lowest setting. Hold the rifle in front of the backsight with the right hand and simultaneously pivot on the left foot to incline the body to the left. Close the right foot to the left foot and recover to the position of attention.

Fig.49 Loading in the Standing Position

# 3. Marksmanship in Supported Positions

## A. Prone Supported Position

### 1. Holding
Rest the rifle hand guard on the support. Hold the rifle with the left hand in front of the magazine, with the back of the left hand resting against the support (or rest the hand on top of the support) and the elbows apart (Fig.50). Alternatively, rest the hand guard directly on the support and hold the butt in the right shoulder with the left hand (Fig.51). Hold the hand grip with the right hand and position the first joint of the index finger on the trigger. The upper right arm should be approaching the vertical. Hold the rifle firmly back into the shoulder with both hands. The head should be slightly forward and the cheek should naturally rest on the butt.

Fig.50 Prone Supported Position - Left Hand in Front of the Magazine

Fig.51 Alternative Prone Supported Position - Left Hand on the Butt

The position of the rifle in the shoulder must be correct. If the rifle is too high in the shoulder, it will tend to shoot with a high point of impact. If it is too low in the shoulder, it will tend to shoot with a low point of impact. The rifle should also be held straight and not canted. If the rifle is canted to the left, it will shoot with the point of impact to the left. If the rifle is canted to the right, it will shoot with the point of impact to the right (Fig.52).

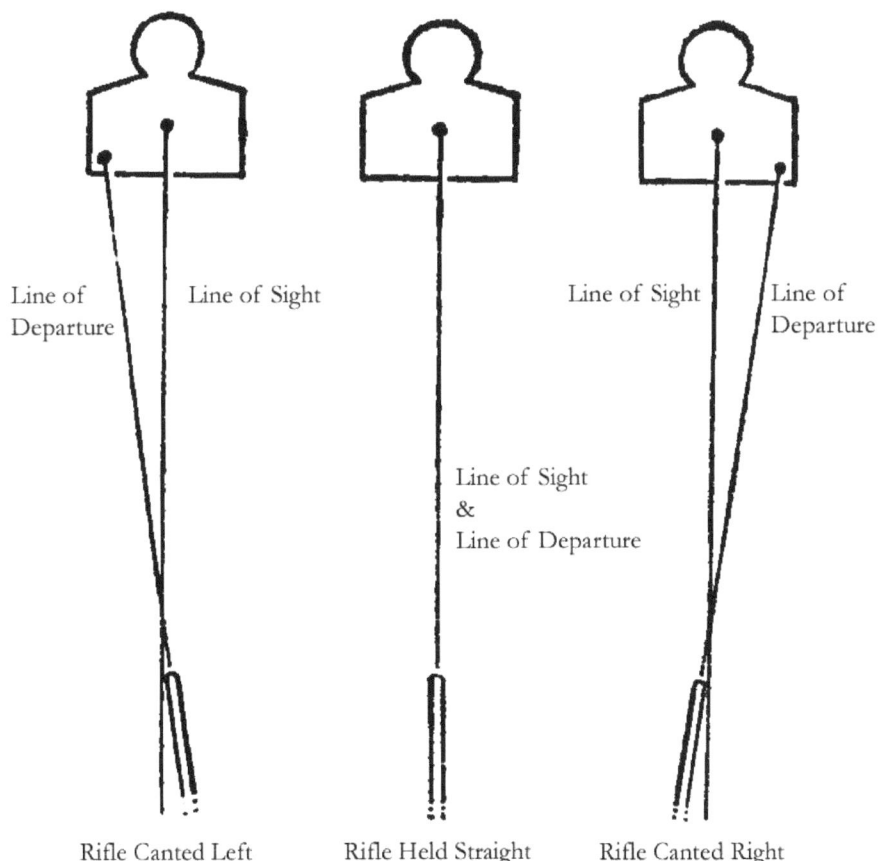

Fig.52 Effect of Canting the Rifle

## 2. Aiming

Look through the backsight notch (or aperture) with the right eye to the foresight and align the sights with the point of aim. The foresight should be aligned in the centre of the sight notch and the foresight tip should be level with the upper edge of the notch (or in the centre of the aperture), and aligned with the point of aim (Fig.53 & Fig.54). If the line of sight is not aligned with the point of aim, move the body to make the correction. If the height requires adjustment, adjust the upper or the lower body, or alternatively open or close the elbows.

Fig.53 Correct Sight Alignment of the Foresight and Backsight

Fig.54 Correct Sight Picture on the Point of Aim

The most important aspect of aiming is the correct sight picture with the alignment of the foresight and the backsight. If attention is paid only to the foresight and the point of aim, then the correct sight alignment with the backsight will be ignored, resulting in an aiming error when the shot is fired. The correct sight picture (Fig.54) will be acquired when the foresight and the backsight alignment is clearly visible and the target is comparatively blurred.

An incorrect sight picture will have a significant effect on accuracy (Fig.55). For example with the Type 53 Carbine: an error of 1mm alignment of the foresight in the backsight notch will result in an error of 21cm at 100m range. For every additional 100m, the size of the error will double, as the degree of error increases with range.

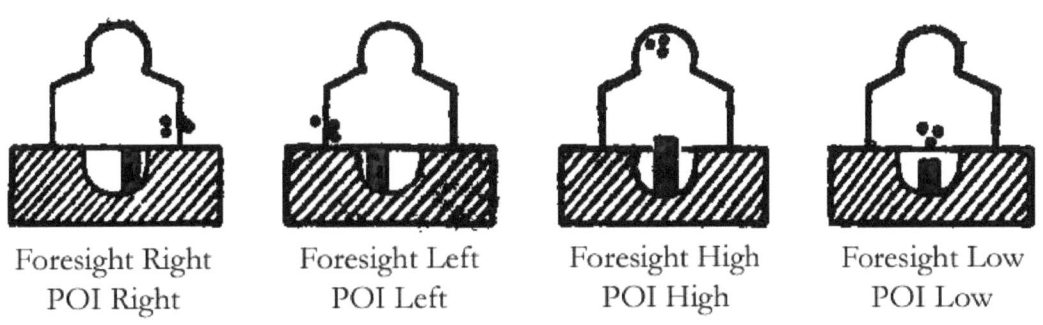

Foresight Right　　Foresight Left　　Foresight High　　Foresight Low
POI Right　　　　POI Left　　　　POI High　　　　POI Low

Fig.55  Effect of Sight Alignment on the Point of Impact

## 3. Firing

Place the first joint of the index finger of the right hand on the trigger. Do not press the trigger. When the point of aim begins aligning in the sight picture, begin firming up even pressure on the trigger and begin slow breathing. When the point of aim is correctly aligned in the sight picture, pause the breathing (there should be no movement in the sight picture) and smoothly increase pressure on the trigger until the rifle fires. If during the firing process, the point of aim drifts off the sight picture or you cannot continue to hold your breath, quickly correct the aim, or breathe, and continue the process of firing the rifle.

Avoid jerking the trigger, blinking, shrugging or relaxing the shoulders at the moment of firing. These actions will change the correct sight alignment and affect the accuracy of the shot.

## B. Kneeling Supported Position

Fig.56  Kneeling Supported (Weapon Pit) Position

When firing from the kneeling supported position in a weapon pit (Fig.56), typically, kneel on the left knee with the left side of the body leaning against the front of the trench, with the lower right leg vertical or braced straight to the rear and to the right. Support the arms by resting the elbows and hold the rifle into the shoulder with both hands. Aim and fire as ordered.

## C. Standing Supported Position

Fig.57 Standing Supported (Weapon Pit) Position

When firing from the standing supported position in a weapon pit (Fig.57), the left leg should be bent, with the left side of the body leaning against the trench and the right leg should be braced straight to the rear. Support the arms by resting the elbows and hold the rifle into the shoulder with both hands. Aim and fire as ordered.

# 4. Marksmanship in Unsupported Positions

## A. Prone Unsupported Position

When firing from the prone unsupported position (Fig.58), the chest is slightly raised from the ground. Hold the rifle under the backsight with the left hand, with the left elbow directly underneath the rifle as much as possible. The elbows should be apart and the upper and lower right arm should form a 90 degree angle, with the upper arm approaching the vertical. Hold the hand grip with the right hand and hold the rifle into the shoulder with both hands. Aim and fire as ordered.

Fig.58 Prone Unsupported Position

## B. Kneeling Unsupported Position

When firing from the kneeling unsupported position (Fig.59), hold the rifle under the backsight with the left hand. The left elbow should be in front of, or behind the left kneecap. The rifle, lower left arm and lower left leg should form a vertical line, and the upper body should lean slightly forward. Hold the hand grip with the right hand and hold the rifle into the shoulder with both hands. Aim and fire as ordered.

Fig.59 Kneeling Unsupported Position

## C. Standing Unsupported Position

When firing from the standing unsupported position (Fig.60), hold the rifle under the magazine with the left hand. The upper left arm should be resting tight against the ribs and the lower arm should approach the vertical. Alternatively, hold the lower hand guard with the left hand, without resting the upper left arm against the ribs. Hold the hand grip with the right hand and hold the rifle into the shoulder with both hands. Aim and fire as ordered.

Fig.60 Standing Unsupported Position

# 5. Correcting the Aim

## A. Individual Check

Rest the rifle on a rifle rest and take aim. Move the head up and down slightly to check whether the foresight is correctly centred in the backsight notch (or aperture). Move the head slightly to the left and right to check whether the foresight tip is correctly level with the upper edge of the backsight notch.

As a further step, after taking aim, an assistant places a sheet of white paper in front of the foresight obstructing the view of the target. Without moving the rifle, the firer checks the alignment of the sights. If the sight alignment is incorrect, realign the backsight notch (or aperture) and foresight, by moving only the head and not the rifle. Remove the sheet of paper and look through the sights towards the target, to determine if the alignment of the sights on the point of aim is correct. This method can be repeated until the correct sight picture on the point of aim is established.

## B. Four Point Aim Check

Rest the rifle on a rifle rest and secure a sheet of white paper 10m in front of the rifle (Fig.61). The assistant holds the aiming disc to the paper target and the instructor takes aim at the black spot of the aiming disc with the rifle, to establish the 'correct point of aim'. The assistant punches a hole in the paper target through the hole in the centre of the aiming disc (Fig.62) with a needle or pencil, to establish the 'correct point of impact'. The firer views the sight picture without moving the instructor's aim and notes the correct sight alignment and sight picture on the 'correct point of aim'. The assistant removes the aiming disc and marks the 'correct point of impact' with an 'X'. The assistant then returns the aiming disc to the target in the approximate location. Continuing to look through the sights, the firer then orders the assistant to move the aiming disc back to the 'correct point of aim', with the commands "UP", "DOWN", 'LEFT" and "RIGHT". When the firer judges that the aiming disc target has returned to the 'correct point of aim' in the sight picture, he gives the command "STOP". The assistant marks the position of the firer's 'indicated point of aim' by making a hole in the target, through the hole in the centre of the aiming disc. This is the firer's 'indicated point of impact'.

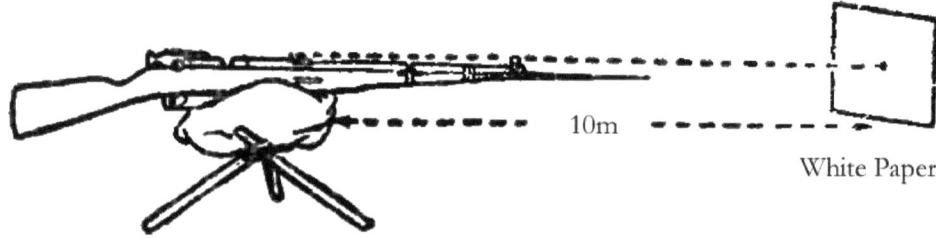

Fig. 61 Four Point Aim Check

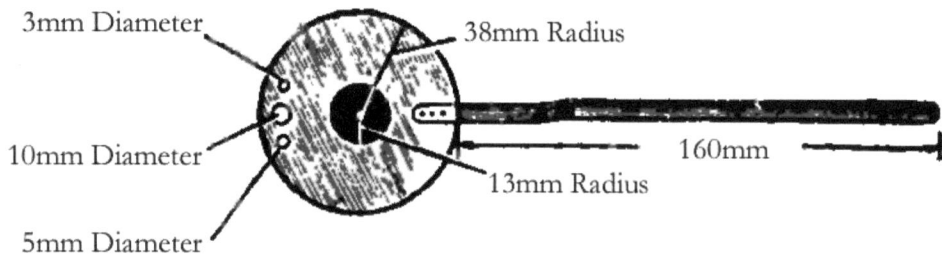

Fig.62 Four Point Aiming Disc

The firer repeats this process three times and the Mean Point of Impact of this 'group' should not exceed 5mm distance from the 'correct point of impact'. The firer's three 'indicated points of impact' should group within a 10mm diameter to 'qualify'. A group within a 5mm diameter is considered 'good' and a group within a 3mm diameter is considered 'excellent'. Since only the target and not the rifle is moved in the Four Point Aim Check technique, elevation and windage errors in the firer's 'indicated points of aim' can be clearly contrasted against the instructor's 'correct point of aim'.

## C. Aim Corrector

Attach the aim corrector (Fig.63) to the rifle, behind the backsight. The instructor views the sight picture through the reflective glass from the left hand side of the rifle. The aim corrector allows the errors in the firer's aiming and firing techniques to be instantly seen and corrected.

Fig.63 Aim Corrector (Universal)[11]

Fig.64 Instructing with the Aim Corrector

---

[11] Translator's Note: The Aim Corrector (Universal), was used for the Type 53 Carbine, Type 56 (SKS) Carbine, Type 56 (AK-47) and Type 63 Assault Rifles. For an example, see 'Militia Weapons and Accessories for the Collector'.

# Submachine Guns

## 1. Safety Inspection

For the purposes of safety, submachine guns must be inspected before and after use, and whenever else it is required. The chamber and magazine must be cleared of ammunition. During the inspection, pointing the muzzle at another person is strictly prohibited.

On the command "SAFETY INSPECTION", grasp the hand guard with the right hand, pointing the muzzle to the front. Unsling the submachine gun from the shoulder and simultaneously incline the body to the right, pivoting on the right foot and step the left foot one pace forward (one shoulder width). Extend the submachine gun forward with the right hand and hold the rear of the hand guard with the left hand (first unfold the stock of the Type 54 Submachine Gun). The upper left arm is held close to the left ribs, the butt (or stock) held against the hip and the foresight around shoulder height. Depress the magazine catch with the right thumb, remove the magazine and place it in the left hand. Release the safety catch, grasp the cocking handle and look to the front. When the commander inspects the submachine gun, retract the cocking handle. After inspection, in your own time, return the bolt forward, apply the safety catch, insert the magazine and hold the hand grip (or pistol grip).

On the command "SAFETY INSPECTION COMPLETE", return the left hand to the hand guard (first fold the stock of the Type 54 Submachine Gun) and move the right hand to the sling. Simultaneously sling the submachine gun onto the right shoulder and incline the body to the left. Close the right foot to the left foot and recover to the position of attention.

## 2. Loading and Sight Setting

### A. Filling Magazines

Hold the magazine with the left hand, with the magazine platform facing upward and the rear of the magazine facing the left. Holding cartridges in the right hand, push each round into the magazine with the base of the cartridge facing the left (the rear of the magazine), with the thumbs of both hands (Fig. 65).

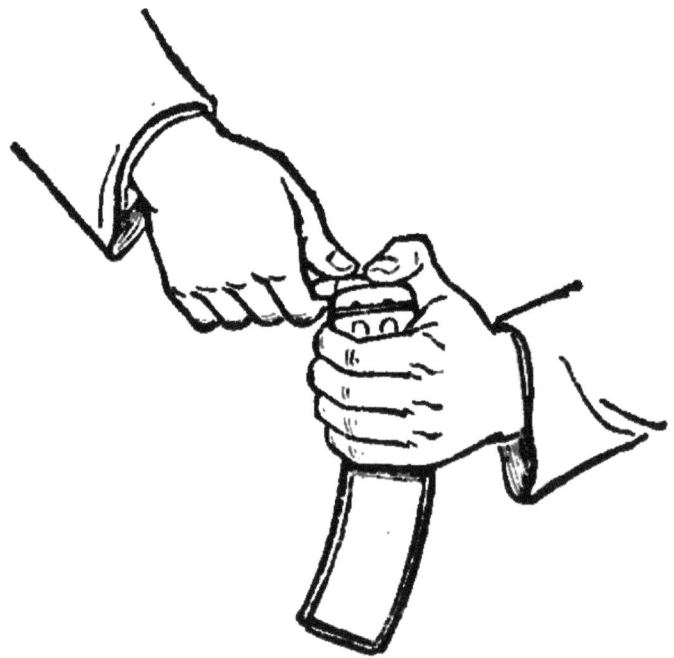

Fig.65 Filling the Magazine

## B. Prone Position

On the command "PRONE - LOAD", hold the hand guard with the right hand, point the muzzle to the front and unsling the submachine gun from the shoulder. Step the left foot one pace forward, stretch the left arm out in the direction of the left foot with palm down and fingers outstretched slightly to the right. Quickly go down on the left knee and lay on the ground, resting on the left hand side of the body supported by the left elbow. Point the submachine gun towards the target with the right hand and hold the rear of the hand guard with the left hand (first unfold the stock of the Type 54). Lean the submachine gun slightly to the left and rest the butt on the ground. Remove the empty magazine with the right hand, place it in the left hand and insert a loaded magazine (Fig.66). Release the safety catch, retract the bolt fully to the rear and apply the safety catch. Return the empty magazine to the magazine pouch. Set the sight and hold the hand grip (or pistol grip) with the right hand. Spread both legs so that the feet are shoulder width apart and pointed outwards, with the body forming a line on the right side with the submachine gun. Watch the front and prepare to fire.

On the command "UNLOAD - STAND", lean the body slightly to the left, remove the magazine with the right hand and place it in the left hand. Release the safety catch, ease the bolt forward and apply the safety catch. Place the empty magazine into the magazine pouch and return the sight to the lowest setting (fold the stock of the Type 54). Grasp the hand guard with the right hand and draw the submachine gun backwards. At the same time, turn the body to the right, allow the left leg to naturally bend back, and draw the left forearm in slightly. Using the left hand, knee and leg for support, stand up and simultaneously step one large pace forward with the right foot. Close the left foot with the right foot. Hold the hand guard with the left hand and simultaneously move the right hand to the sling. Sling the submachine gun on the right shoulder and recover to the position of attention.

Fig.66 Loading in the Prone Position

## C. Kneeling Position

On the command "KNEEL – LOAD", hold the hand guard with the right hand with the muzzle pointing to the target and unsling the submachine gun from the shoulder. Step forward one pace with the left leg, extend the submachine gun towards the target with the right hand and simultaneously hold the rear of the hand guard with the left hand. Kneel down on the right knee and sit on the heel of the right foot. The lower left leg should be nearly vertical and both legs should be held at close to a 90 degree angle. Rest the left forearm on the left thigh (first unfold the stock of the Type 54). Load the submachine gun, set the sight and hold the hand grip (or pistol grip) with the right hand. Watch the front and prepare to fire.

On the command "UNLOAD – STAND", unload the submachine gun and return the sight to the lowest setting (fold the stock of the Type 54). Stand up, move the right hand to the sling and sling the submachine gun on the right shoulder. At the same, close both feet together and recover to the position of attention.

## D. Standing Position

On the command "STAND – LOAD", hold the hand guard with the right hand, with the muzzle pointing to the target and unsling the submachine gun from the shoulder. Incline the body to the right, pivoting on the right foot and step the left foot one pace forward (one shoulder width). Balance the body's weight equally on both feet. Extend the submachine gun forward with the right hand and hold the rear of the hand guard with the left hand (unfold the stock of the Type 54). The left upper arm should be held close to the left ribs and the butt (or stock) held near the hip. The muzzle should be held at shoulder height. Load the submachine gun, set the sight and hold the hand grip (or pistol grip) with the right hand. Watch the front and prepare to fire.

On the command "UNLOAD", unload the submachine gun and return the sight to the lowest position (fold the stock of the Type 54). Pivot on the left foot to incline the body to the left and close the right foot to the left foot. Move the right hand to the sling and sling the submachine gun on the right shoulder. Recover to the position of attention.

# 3. Marksmanship in Supported Positions

## A. Prone Supported Position

### 1. Holding

Rest the hand guard on the support and hold the magazine or the front of the stock with the left hand (or the magazine housing of the Type 54) with the elbows apart (Fig.67). Hold the hand grip (or pistol grip) with the right hand and position the first joint of the index finger on the trigger. The upper right arm should approach the vertical. Hold the submachine gun firmly back into the shoulder with both hands. The head should be slightly forward and the cheek should naturally rest on the butt (or stock).

Left Hand Holding the Magazine

Left Hand Holding the Stock

Left Hand Holding the Magazine Housing (Type 54)

Fig.67 Alternative Prone Supported Positions

The position of the submachine gun in the shoulder must be correct. If the submachine gun is too high in the shoulder, it will tend to shoot with a high point of impact. If it is too low in the shoulder, it will tend to shoot with a low point of impact. The submachine gun should also be held straight and not canted. If the submachine gun is canted to the left, it will shoot with the point of impact to the left and if canted to the right, it will shoot with the point of impact to the right (Fig.68).

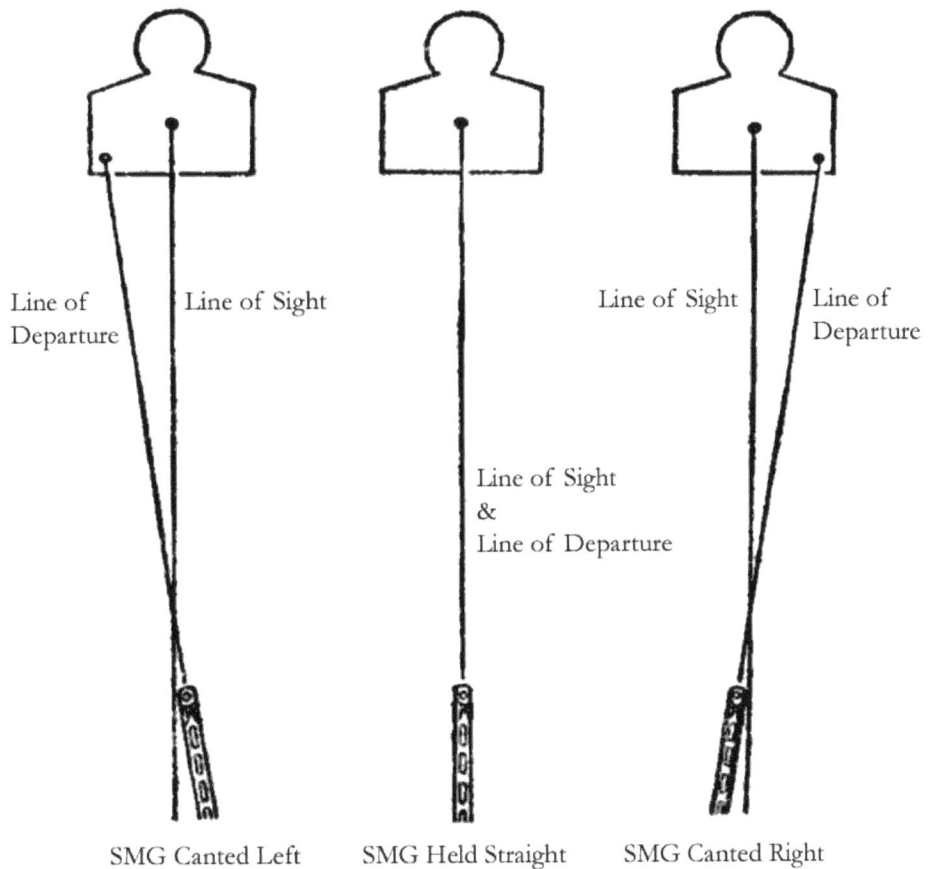

Fig.68 Effect of Canting the Submachine Gun

## 2. Aiming

Look through the backsight aperture (or notch) with the right eye to the foresight and align the sights with the point of aim. The foresight should be aligned in the centre of the aperture (or in the centre of the sight notch and the foresight tip level with the upper edge of the notch), and aligned with the point of aim (Fig.69 & Fig.70). If the line of sight is not aligned with the point of aim, move the body to make the correction. If the height requires adjustment, adjust the front or the back of the body, or alternatively open or close the elbows.

Fig.69 Correct Sight Alignment of the Foresight and Backsight

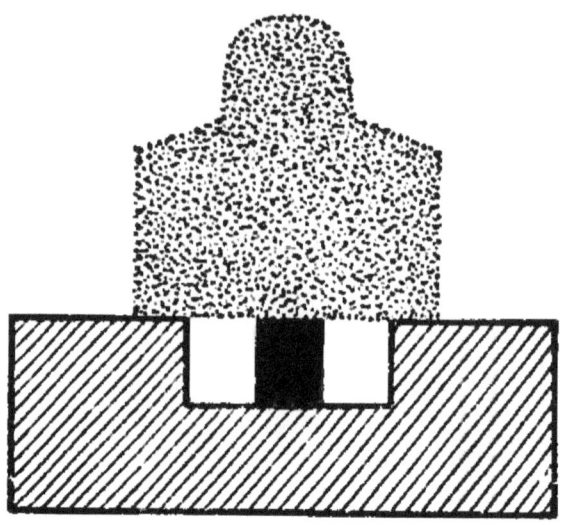

Fig.70 Correct Sight Picture on the Point of Aim

The most important aspect of aiming is the correct sight picture with the alignment of the foresight and the backsight. If attention is paid only to the foresight and the point of aim, then the correct sight alignment with the backsight will be ignored, resulting in an aiming error when the shot is fired. The correct sight picture (Fig.70) will be acquired when the foresight and the backsight alignment is clearly visible and the target is comparatively blurred.

An incorrect sight picture will have a significant effect on accuracy (Fig.71). For example with the Type 50 Submachine Gun: an error of 1mm alignment of the foresight in the backsight aperture will result in an error of 26cm at 100m range. For every additional 100m, the size of the error will double, as the degree of error increases with range.

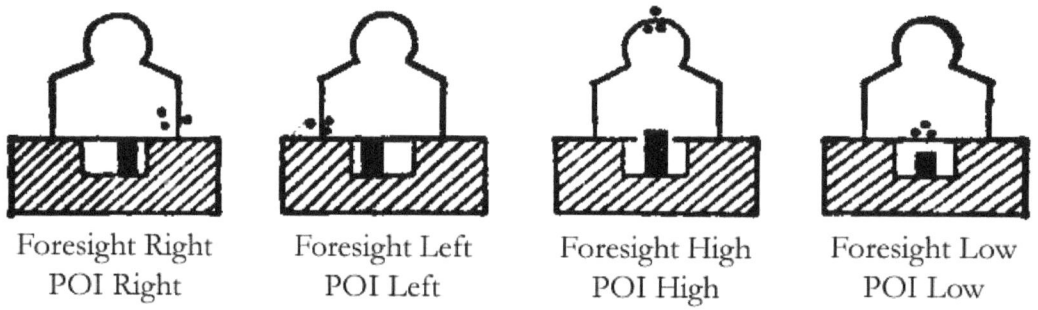

Fig.71 Effect of Sight Alignment on the Point of Impact

## 3. Firing

Place the first joint of the index finger of the right hand on the trigger. Do not press the trigger. When the point of aim begins aligning in the sight picture, begin firming up even pressure on the trigger and begin slow breathing. When the point of aim is correctly aligned in the sight picture, pause the breathing (there should be no movement in the sight picture) and smoothly increase pressure on the trigger until the submachine gun fires. If during the firing process, the point of aim drifts off the sight picture or you cannot continue to hold your breath, quickly correct the aim, or breathe, and continue the process of firing the submachine gun.

Avoid jerking the trigger, blinking, shrugging or relaxing the shoulders at the moment of firing. These actions will change the correct sight alignment and affect the accuracy of the shot.

## B. Kneeling Supported Position

Fig.72 Kneeling Supported (Weapon Pit) Position

When firing from the kneeling supported position in a weapon pit (Fig.72), typically, kneel on the left knee with the left side of the body leaning against the front of the trench, with the lower right leg vertical or braced straight to the rear and to the right. Support the arms by resting the elbows and hold the submachine gun into the shoulder with both hands. Aim and fire as ordered.

## C. Standing Supported Position

Fig.73 Standing Supported (Weapon Pit) Position

When firing from the standing supported position in a weapon pit (Fig.73), the left leg should be bent, with the left side of the body leaning against the trench and the right leg should be braced straight to the rear. Support the arms by resting the elbows and hold the submachine gun into the shoulder with both hands. Aim and fire as ordered.

# 4. Marksmanship in Unsupported Positions

## A. Prone Unsupported Position
When firing from the prone unsupported position (Fig.74), the chest is slightly raised from the ground. Hold the magazine or the front of the stock (or the magazine housing of the Type 54) with the left hand, with the left elbow directly underneath the submachine gun as much as possible. The elbows should be apart, and the left hand should hold the submachine gun down with force. Hold the hand grip (or pistol grip) with the right hand and hold the submachine gun into the shoulder with both hands. Aim and fire as ordered.

Fig.74 Prone Unsupported Position

## B. Kneeling Unsupported Position
When firing from the kneeling unsupported position (Fig.75), hold the magazine or the front of the stock (or the magazine housing of the Type 54) with the left hand. The left elbow should be in front or behind the left kneecap. The submachine gun, lower left arm and lower left leg should form a vertical line, and the upper body should lean slightly forward. Hold the hand grip (or pistol grip) with the right hand and hold the submachine gun into the shoulder with both hands. Aim and fire as ordered.

Fig.75 Kneeling Unsupported Position

## C. Standing Unsupported Position

When firing from the standing unsupported position (Fig.76), hold the magazine or the front of the stock (or the magazine housing of the Type 54) with the left hand. The upper left arm should be resting tight against the ribs and the lower arm should approach the vertical. Hold the hand grip (or pistol grip) with the right hand and hold the submachine gun into the shoulder with both hands. Aim and fire as ordered.

Fig.76 Standing Unsupported Position

# Marksmanship under Combat Conditions

## 1. Firing at Fleeting Targets

Fleeting targets are difficult to see, as they may appear with short exposure times and in unpredictable locations. Therefore, fleeting targets must be spotted quickly. Sight setting, aiming and firing must be quick and decisive. If a target is concealed and it is not possible to determine its location, prepare to aim and fire again to destroy it when it reappears. Fleeting targets can be engaged by:

1. Deliberate Shot. Determine the pattern of the target's movements and aim where the target is expected to reappear. When the target does reappear, quickly correct the aim and fire.

2. Snap Shot. When it is not possible to determine where the target may reappear, aim in the general direction of the target and when the target reappears, quickly fire a snap shot.

## 2. Firing at Moving Ground Targets

Moving targets have constantly changing positions, directions, speed and range. Firing on a direct aim taken at a moving target will result in a miss, since the target will have already moved by the time that the shot has been fired and the bullet arrives at the point of aim. Therefore, it is necessary to make an allowance by aiming off and firing ahead of the target.

### A. Crossing Target Allowance

The calculation for the allowance required for a crossing target is:

Moving target speed (m/s) x bullet flight time (s)

**Table: Moving Target Speeds (m/s)**

| Type of Target | Type of Movement | | | | | |
|---|---|---|---|---|---|---|
| | Walking | Jogging | Running | Slow | Moderate | Fast |
| Infantry | 1.5 | 3 | 4.5 | - | - | - |
| Cavalry | 2 | 4 | 8 | - | - | - |
| Motor Vehicles | - | - | - | 4 | 8 | 12 |

**Table: Bullet Flight Time (s)**

| Weapon | Range (m) | | | |
|---|---|---|---|---|
| | 100 | 200 | 300 | 400 |
| Rifle | 0.13 | 0.28 | 0.41 | 0.62 |
| Submachine Gun | 0.23 | 0.54 | 0.89 | - |

Example: If firing a rifle at infantry jogging across the front at 200m, the crossing target allowance is: 3 m/s x 0.28 s = 0.84m. More conveniently, the allowance can be measured in terms of body widths (0.4m).

Accordingly, the amount of allowance measured in body widths, for firing a rifle at infantry jogging (3 m/s) across the front at 200m, can be taken from the following table:

## Table: Rifle - Crossing Target Allowance for Jogging Infantry (Body Widths)

| Range (m) | 100 | 200 | 300 | 400 |
|---|---|---|---|---|
| Allowance | 1 | 2 | 3 | 4 |

When firing at infantry jogging obliquely across the front, the allowance is halved (at 200m range, the allowance is 1 body width).

The amount of allowance measured in body widths, for firing a submachine gun at infantry jogging (3 m/s) across the front at 200m, can be taken from the following table:

## Table: Submachine Gun - Crossing Target Allowance for Jogging Infantry (Body Widths)

| Direction of Target Movement | Range (m) | |
|---|---|---|
|  | 100 | 200 |
| Crossing | 1.5 | 3 |
| Oblique | 1 | 2 |

When firing at crossing or oblique targets moving into a headwind, reduce the amount of allowance. When firing at crossing or oblique targets moving with a tailwind, increase the allowance. The amount of allowance should be taken from the centre of the target (Fig.77).

Fig.77 Crossing Target Allowance of One Body Width

## B. Engaging Crossing Targets

### 1. Ambush Shot
Select a point of aim by aiming off in front of a moving target and prepare to fire. When the target approaches the point of aim, quickly correct the elevation and gradually increase pressure on the trigger. When the target has moved into the correct sight picture, fire decisively, without hesitation or snatching of the trigger.

### 2. Tracking Shot
Align the sights in the direction of the moving target. Aim at the target (aim off allowing for wind as required) and swing the rifle evenly, tracking the movement of the target. Maintain the allowance (if any) and the correct sight picture, as well as an even pressure on the trigger. Fire at the target when the sight picture is correct, maintaining the swing tracking of the target. Do not track the target for too great a distance.

When firing at a target moving towards or away from the firer once the sight has been set, adjust the point of aim by aiming off. For a target moving towards the firer, aim at the lower part of the body. For a target moving away from the firer, aim at the upper part of the body.

## 3. Firing in Mountainous Terrain

Mountainous areas have complex terrains and greatly variable weather conditions, making observation and ranging difficult. Firing positions can be unstable and the relative positions of the firer and targets are typically not level. As a result, it is easy for a shot to miss high or long, regardless of whether the firer is firing uphill or downhill. Therefore when firing in mountainous terrain, choose a stable firing position with a clear field of view and a wide field of fire. Observe the target carefully to accurately determine the range, and reduce the sight setting or lower the point of aim as required.

When firing uphill, take a firing position as high as possible. When firing from the prone position, the feet should be wider apart compared to when firing on level ground, with the feet dug into the ground or braced against an object. When necessary, bend the right leg so that the body is closer to the ground, bearing the weight on the left foot and bracing the right foot to the rear.

When firing downhill, any firing rest used should not be too high. When firing from the prone position, the abdomen should be distributed evenly on the ground and the feet should be wider apart compared to when firing on level ground. The left hand supporting the weapon should be slightly forward and the right elbow should raise the height of the shoulder (submachine guns can use the base of the magazine for support on the ground). When firing in the kneeling or standing positions, adopt an appropriately stable position.

When firing at a target within 300m range at an angle of 20 degrees or less, there is generally no need to make an allowance for aim. If the range or angle of the target is greater than 300m or 20 degrees, lower the sight setting or lower the point of aim (see the Table below). When firing at a target moving uphill, aim at the upper half of the body; for a target moving downhill, aim at the lower half of the body. When firing at crossing and obliquely moving targets, make a crossing target allowance for aim.

Table: Allowance for Range and Angle in Mountainous Terrain - Point of Aim (cm)

| Angle of Target (degrees) | Sloping Range (m) | | | |
|---|---|---|---|---|
| | 100 | 200 | 300 | 400 |
| 20 | -1 | -3 | -6 | -10 |
| 25 | -2 | -4 | -9 | -14 |
| 30 | -3 | -5 | -11 | -18 |
| 35 | -4 | -5 | -13 | -24 |

## 4. Firing at Airborne Targets

### A. Firing at Aircraft

Aircraft are fast moving and change direction and altitude rapidly. Therefore in order to destroy enemy aircraft, there should be a rapid engagement of the aircraft with concentrated fire by the squad (or platoon) under the control of the commander (Fig.78). Aircraft are generally engaged with small arms within a range of 500m, with the sight set at '3'.

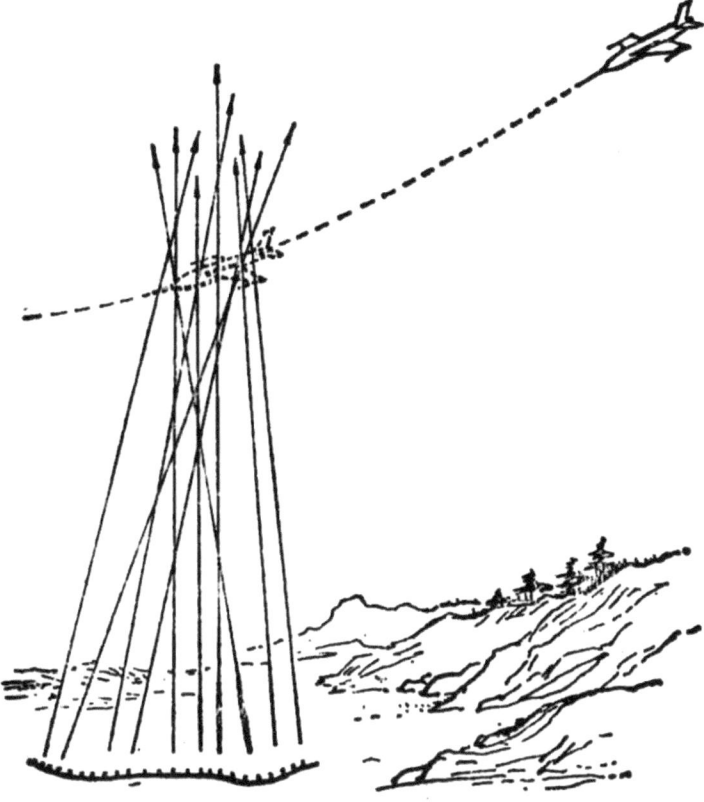

Fig.78  Concentrated Fire

The most effective time to fire at an enemy aircraft, is when it is diving towards or climbing away from the firer. Aim at the nose of a diving aircraft and at the tail of a climbing aircraft (Fig.79). The most effective time to fire at a helicopter is when it is hovering, and when it is climbing or descending vertically. Aim at the middle of the helicopter when it is hovering. When it is climbing, aim at the upper part of the helicopter and when it is descending, aim at the lower part of the helicopter.

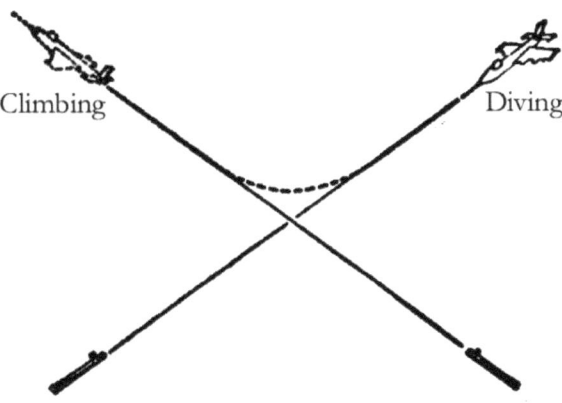

Fig.79  Firing at Climbing and Diving Aircraft

When firing at aircraft flying level or diving at a distant target, aim off with an allowance according to the speed of the aircraft. The allowance is taken from the nose of the aircraft. When firing at aircraft flying at 300 m/s or faster, the allowance is half that of the figure of the range of the aircraft (Fig.80). Additionally the allowance can be determined from the following table.

**Table: Allowance for Firing at Aircraft with Rifles (Aircraft Lengths)**

| Aircraft Speed (m/s) | | Range (m) | | | |
|---|---|---|---|---|---|
| | | 200 | 300 | 400 | 500 |
| 160 | Large Size | 2 | 3 | 4.5 | 6 |
| | Small Size | 3 | 5 | 7 | 9.5 |
| 300 | Large Size | 4 | 6 | 8.5 | 11.5 |
| | Small Size | 6 | 10 | 14 | 18 |
| 360 | Large Size | 4.5 | 7.5 | 10 | 13.5 |
| | Small Size | 7.5 | 12 | 16.5 | 22 |
| 500 | Large Size | 6.5 | 10 | 14 | 18.5 |
| | Small Size | 10 | 16 | 22.5 | 30 |
| 62 | Helicopter | 1 | 2 | 3 | 3.5 |
| Definition | | A large aircraft has an average length of 21m. A small aircraft has an average length of 13m. | | | |

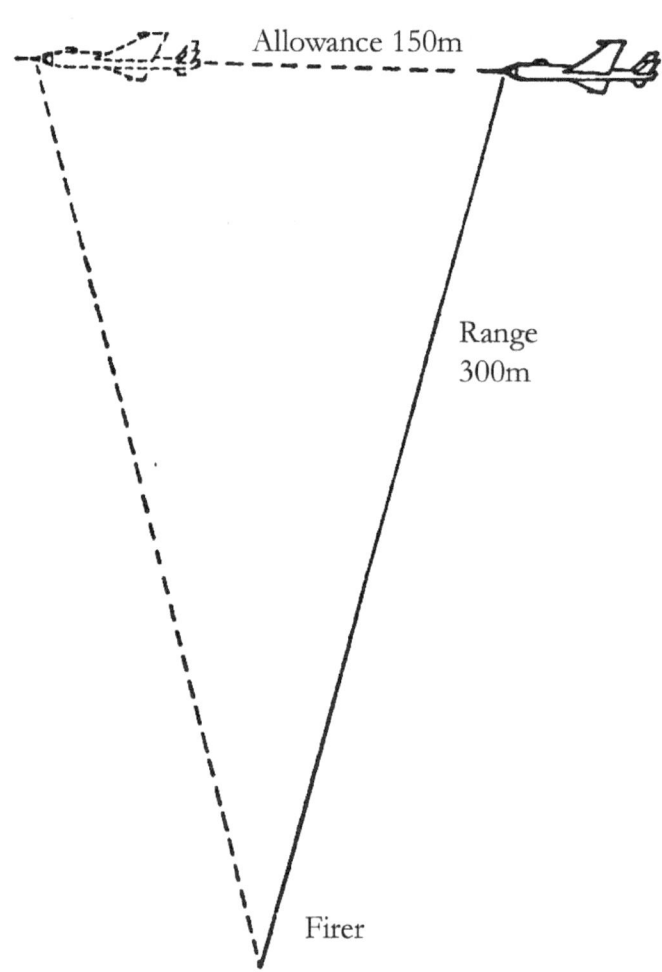

Fig.80 Allowance is Half of the Figure of the Range

## B. Firing at Paratroopers

Paratroopers have a slow rate of descent in dense and randomly distributed formations. When firing at descending paratroopers, generally fire first on the nearer and lower troops before the farther and higher troops, and fire first on officers (different coloured parachutes) before troopers. Attempt to destroy the enemy in mid-air. When firing at paratroopers within 500m range, set the sight at '3' with a point of aim or allowance indicated in the following Table. The allowance should be in the direction of the paratrooper's descent, determined from a point of aim at the paratrooper's feet (Fig.81).

**Table: Allowance for Firing at Descending Paratroopers (Body Lengths)**

| Range (m) | 100 | 200 | 300 | 400 | 500 |
|---|---|---|---|---|---|
| Rifle | 0 | 1 | 1.5 | 2.5 | 3.5 |
| Submachine Gun | 0.5 | 1.5 | 3 | | |

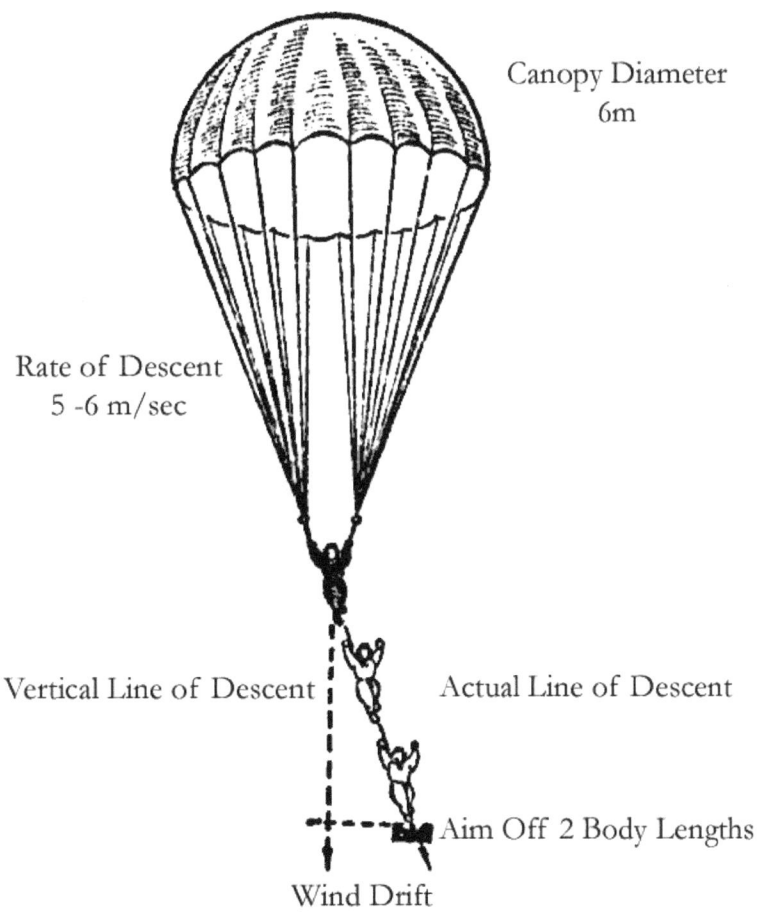

Fig.81 Aiming Off When Firing at Paratroopers

## C. Positions for Firing at Airborne Targets

When firing at airborne targets, take any position that is appropriate to the nature of the enemy target and the terrain. Use objects as weapon rests where possible. When there are no rests available, take any position, be it standing, kneeling, sitting or lying on the back (Fig.82).

Fig.82 Positions for Firing at Airborne Targets

# 5. Firing at Seaborne Targets

The ocean has wide fields of view and clear fields of fire. However it is easy to make errors in determining range. Targets are subject to the effects of winds, waves, tides and currents, and their constant vertical and horizontal motion makes holding and aiming difficult. Great variations in the weather, rising mist, reflections and glare from the water, all affect aiming and errors are easily made. However, seaborne targets are subject to constant wave motions which take a regular pattern, and targets rise and fall with the wave motion.

When firing at seaborne targets, set the sight and aim at the centre or the lower half of the target. Ensure that the sight picture is correct and take up pressure on the trigger. Targets may be fired on accurately by using the regular patterns of wave motions. Fire when the target is high on the crest of a wave or in the trough prior to the next rising wave. When firing across the direction of the wave motion (when the wave motion is rocking the target from side to side), fire when the target is between two waves. In order to obtain the best results, ensure that the hold is correct, the aim is rapidly acquired and the shot is fired quickly.

When firing at seaborne targets, an angle of depression is formed between the target and the firer. If the angle of depression is 25 degrees or less and the range does not exceed 400m, there is generally no need for an allowance on the aim.

# 6. Firing at Night

When firing at night, the limited visibility makes it relatively difficult to spot targets, set sights and to aim. Firing at night is generally conducted in low light conditions such as flare illumination or moonlight, and by firing in fixed lines.

## A. Firing in Flare Illumination

Firing in flare illumination is generally conducted with a predetermined sight setting. When firing in flare illumination, quickly adopt a firing position, point the weapon at the target and aim by the illumination of the flare. The procedure for aiming is: find the source of the illumination, find the foresight, and then find the backsight. Use the foresight guard to frame the illumination, and then wiggle the rifle until the foresight is seen. Slightly lower the foresight to find the backsight notch (or aperture), then realign the sights. If the line of sight is directly towards the flare illumination, fire decisively. If the flare illumination is behind the line of sight and the illumination is lost, maintain the firing position and quickly fire the shot.

If the enemy is firing from the side, point the muzzle towards the muzzle flashes to determine the enemy's position. Aim at the enemy's silhouette and fire.

Aiming at night is difficult, and errors in aiming should be noted and corrected. Typically, when the foresight can be clearly seen, then the foresight tip is higher than the backsight notch. Conversely, when the foresight cannot be seen, then the foresight tip is lower than the backsight notch. If the foresight cannot be seen, wiggle the rifle slightly until it is visible.

## B. Firing in Moonlight

When firing in moonlight and the target can be clearly seen, aim directly and fire. If the target is indistinct, fire at the silhouette. Use the relatively brighter background of the area around the target to obtain a correct sight alignment, and then shift the sight picture to the target silhouette and fire.

## C. Firing in Fixed Lines

For firing at night in fixed lines, it is necessary to prepare the defences by day (Fig.83). When preparing fixed lines, first determine the desired direction and the field of fire, and accurately measure the range. Prepare two forked tree branches or wooden boards with cut notches, in which the weapon can be rested. Stake these into the parapet of the weapon pit in the defensive position. Rest the hand guard of the weapon in the front support, and rest the front of the trigger guard (or butt stock of a submachine gun) in the rear support. Set the exact range in the sight and align the field of fire. If several points of aim are required, use a rear support with more than one notch cut, or a wider support which allows a degree of lateral movement. If there is insufficient material to prepare a fixed line for a rifle, a channel aligned in the desired direction can be dug into the parapet of the weapon pit, to enable firing in a fixed line.

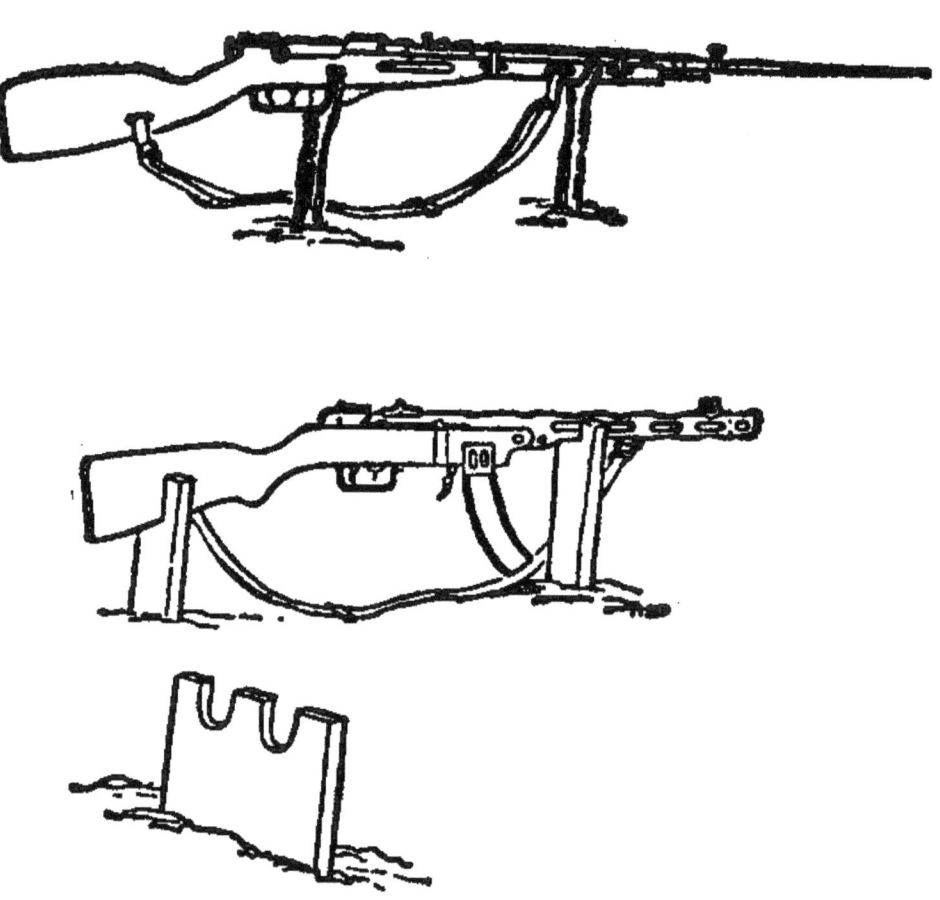

Fig.83 Firing in Fixed Lines

# Appendices

## 1. Rifle Data

| Rifle | Type 53 Carbine (Mosin Nagant) | 79 Rifle (Mauser) | 65 Rifle (Arisaka) | Type 99 Rifle (Arisaka) | 30 Rifle (Springfield) |
|---|---|---|---|---|---|
| Calibre (mm) | 7.62 | 7.9 | 6.5 | 7.7 | 7.62 |
| Weight (kg) | 3.9 | 4.08 | 3.9 | 4.1 | 3.95 |
| Length (mm) | 1020 | 1180 | 1280 | 1258 | 1155 |
| Barrel Length (mm) | 520 | 594 | 769 | 797 | 610 |
| Muzzle Velocity (m/s) | 820 | 810 | 762 | 740 | 807 |
| Sight Radius (mm) | 416 | 504 | 685 | 505 | 562 |
| Rifling – Right Hand (grooves) | 4 | 4 | 4 | 4 | 4 |
| Maximum Sight Range (m) | 1000 | 2000 | 2400 | 1500 | 2700 |
| Projectile Maximum Range (m) | 3000 | 3000 | 3000 | 3000 | 3000 |
| Rate of Fire (rnds/min) | 8 - 10 | 8 - 10 | 8 - 10 | 8 - 10 | 8 - 10 |
| Magazine Capacity (rnds) | 5 | 5 | 5 | 5 | 5 |

## 2. Submachine Gun Data

| Submachine Gun | Type 50 Submachine Gun (PPSh-41) | Type 54 Submachine Gun (PPS-43) |
|---|---|---|
| Calibre (mm) | 7.62 | 7.62 |
| Weight with magazine (kg) | 4.31 | 3.65 |
| Length (mm) | 850 | Stock Unfolded 920 Stock Folded 618 |
| Muzzle Velocity (m/s) | 460 | 500 |
| Sight Radius (mm) | 365 | 352 |
| Rifling – Right Hand (grooves) | 4 | 4 |
| Maximum Sight Range (m) | 200 | 200 |
| Projectile Maximum Range (m) | 800 | 800 |
| Rate of Fire (rnds/min) | 70 – 100 | 70 – 100 |
| Magazine Capacity (rnds) | 30 | 35 |

## 3. Rifles - Height of Point of Impact above the Point of Aim

Height of Point Of Impact above the Point Of Aim (cm)

| Rifle | Sight Setting | Range (m) | | | | |
|---|---|---|---|---|---|---|
| | | 100 | 200 | 300 | 400 | 500 |
| Type 53 Carbine (Mosin Nagant) | 1 | 0 | | | | |
| | 2 | 8 | 0 | | | |
| | 3 | 19 | 23 | 0 | | |
| | 4 | 33 | 51 | 43 | 0 | |
| | 5 | 50 | 80 | 90 | 60 | 0 |
| 65 Rifle (Arisaka) | 1 | 0 | -20 | | | |
| | 2 | 10 | 0 | -34 | | |
| | 3 | 22 | 23 | 0 | -61 | |
| | 4 | 35 | 49 | 39 | 0 | -74 |
| | 5 | 50 | 79 | 84 | 60 | 0 |
| 79 Rifle (Mauser) | 1 | 0 | -21 | | | |
| | 2 | 11 | 0 | -34 | | |
| | 3 | 22 | 32 | 0 | -50 | |
| | 4 | 35 | 49 | 39 | 0 | -73 |
| | 5 | 49 | 70 | 80 | 58 | 0 |
| 30 Rifle (Springfield) | 1 | 0 | | | | |
| | 2 | 7 | 0 | | | |
| | 3 | 18 | 19 | 0 | | |
| | 4 | 28 | 40 | 32 | 0 | |
| | 5 | 30 | 60 | 70 | 50 | 0 |

## 4. Submachine Guns - Height of Point of Impact above the Point of Aim

Height of Point Of Impact above the Point Of Aim (cm)

| SMG | Sight Setting | Range (m) | | | |
|---|---|---|---|---|---|
| | | 50 | 100 | 150 | 200 |
| Type 50 SMG (PPSh-41) | 10 | 7 | 0 | -25 | -66 |
| | 20 | 24 | 33 | 25 | 0 |
| Type 54 SMG (PPS-43) | 10 | 7 | 0 | -24 | -70 |
| | 20 | 24 | 36 | 29 | 0 |

# 5. Rifles - Aiming Off to Correct the Point of Impact

**Amount to Aim Off from the Initial Point of Aim (mm)**

| Rifle | Amount of Correction from Initial Point Of Impact | | | | POI above POA at 100m Sight Setting '3' |
|---|---|---|---|---|---|
| | 15cm | 20cm | 25cm | 30cm | |
| Type 53 Carbine (Mosin Nagant) | 0.6 | 0.8 | 1.0 | 1.2 | 19cm |
| 79 Rifle (Mauser) | 0.75 | 1.0 | 1.26 | 1.51 | 22cm |
| 65 Rifle (Arisaka) | 1.03 | 1.36 | 1.71 | 2.05 | 22cm |
| 30 Rifle (Springfield) | 0.84 | 1.12 | 1.4 | 1.68 | 18cm |

# 6. Ammunition Data

## A. Type 53 Carbine (Mosin Nagant) Cartridges

| Bullet | Bullet Tip Colour | Designation | Case Type | Case Length (mm) | Application |
|---|---|---|---|---|---|
| Light Ball[12] | None | 7.62 普 | Rimmed | 53 | Anti-personnel |
| Steel Core | Silver | 7.62 钢轻 | Rimmed | 53 | Anti-personnel |
| Tracer | Green | 7.62 曳 | Rimmed | 53 | Aim correction; target indication; signalling |
| Armour Piercing Incendiary | Black & Red | 7.62 穿燃 | Rimmed | 53 | Anti-armour (light); anti-materiel |
| Incendiary | Red | 7.62 试燃 | Rimmed | 53 | Anti-materiel |

## B. 79 Rifle (Mauser) Cartridges

| Bullet | Primer Colour | Case Type | Case Length (mm) | Application |
|---|---|---|---|---|
| Round Nose | Black | Rimless | 57 | Anti-personnel |
| Spitzer | Black | Rimless | 57 | |
| Heavy Spitzer | Green | Rimless | 57 | Anti-armour (light) |
| Armour Piercing | Red | Rimless | 57 | |
| Armour Piercing Tracer | Red | Rimless | 57 | |

---

[12] Translator's Note: This designation applies to both lead core (Type L) and steel core (Type LPS) 7.62x54mmR ball cartridges. See also 'Ammunition' pp78-80 for a description of changes in Type 53 ball cartridge specifications and designations.

## C. 65 Rifle (Arisaka) Cartridges

| Bullet | Case Colours | | Case Type | Case Length (mm) | Application |
|---|---|---|---|---|---|
| | Case Mouth | Primer | | | |
| Round Nose | | | Semi-rimmed | 50.5 | Anti-personnel |
| Spitzer | Pink[13] | | Semi-rimmed | 50.5 | |
| Armour Piercing (new) | Black | | Semi-rimmed | 50.5 | Anti-armour (light) |
| Armour Piercing (old) | | Black | Semi-rimmed | 50.5 | |

## D. Type 99 Rifle (Arisaka) Cartridges

| Bullet | Colours | | Case Type | Case Length (mm) | Application |
|---|---|---|---|---|---|
| | Case Mouth | Primer | | | |
| Spitzer | Pink | Green or None | Rimless | 57.5 | Anti-personnel |
| Armour Piercing | Black (new) | Black (old) | Rimless | 57.5 | Anti-armour (light) |
| Tracer | Green (new) | Green (old) | Rimless | 57.5 | Aim correction; target indication; signalling |
| Incendiary | Red (new) | Red (old) | Rimless | 57.5 | Anti-materiel |

## E. 30 Rifle (Springfield) Cartridges

| Bullet | Colours | | Case Type | Case Length (mm) | Application |
|---|---|---|---|---|---|
| | Primer | Bullet Tip | | | |
| Spitzer | Black | | Rimless | 63 | Anti-personnel |
| Spitzer | | | Rimless | 63 | |
| Heavy Spitzer | Red or None | | Rimless | 63 | |
| M1 Armour Piercing | | Black | Rimless | 63 | Anti-armour (light) |
| M1 Tracer | | Red | Rimless | 63 | Aim correction; target indication; signalling |
| M1 Incendiary | | Blue | Rimless | 63 | Anti-materiel |

---

[13] Translator's Note: 6.5mm Arisaka spitzer rounds with pink case sealant were Japanese manufactured, or manufactured under Japanese control. Chinese Communist manufactured spitzer rounds from 1948 – 1952 typically had green case sealant.

# Militia Weapons and Accessories for the Collector

The long process of standardisation of weapons and ammunition by the Militia which began around the time of the 1960 – 1961 Militia reforms, was finally completed around the time of the 1979 – 1981 reforms. By the early 1980s, the Type 56 (SKS) carbine, the Type 56 (AK-47) and Type 63 assault rifles with the standard calibre of 7.62x39mm had finally replaced the great diversity of bolt action rifles and submachine guns dating to the World War II era.[14]

In the early 1980s, Western military rifle collectors and shooters were the unexpected beneficiaries of China's new economic reforms and 'Open Door Policy'. Eager for foreign trade, China's state arsenals found new customers in the form of Western collectors and shooters, for its vast stocks of Type 56 (SKS) carbines. One observer in the Australian firearms industry reported that the sales negotiated for Type 56 carbines included the bundled disposal of equal or greater quantities of obsolete Chinese military firearms in various conditions, including those bolt action rifles covered in this manual, with each Type 56 carbine sold.[15] Australian and American firearms industry observers reported that the Type 53 carbines along with other obsolete firearms, had been collected and stored in "cave" like warehouses, bundled in groups of 10, then tied with bailing wire and 'tourniqueted' with a stick, with 20 such bundles wired together onto a pallet.[16]

Thousands of such obsolete bolt action rifles were exported from China during this period and acquired by Western collectors. While the Springfields and to a lesser extent the Arisakas were well known to Western collectors, apart from the relatively rare examples of Type 53 carbine or Chiang Kai-shek Mauser 'war trophies' in the United States, these imports in the 1980s were the first glimpse of these types by Western collectors. Some 'excellent' or 'good' examples, especially of Type 53 carbines were found. However, many if not the majority of rifles ranged from 'average' to 'poor' condition, with damaged stocks, worn bores, missing and/or 'mismatched' serial numbered parts. The lack of reliable information about the Chiang Kai-shek Mauser and Type 53 carbine in particular in the early days of collecting these rifles, led to a number of misconceptions amongst collectors. In time, with research involving a wider range of collector's rifles from outside the United States, particularly Europe and Australia, and a better understanding of Chinese history and politics, a number of these misconceptions can be corrected.

While a discussion of each of these firearms types is outside of the scope of this book, a number of points of interest to collectors can be made. The varying conditions of these rifles, some in very poor cosmetic or mechanical condition, some damaged, some poorly maintained, some with missing parts etc, led to some erroneous beliefs amongst some collectors about "Chinese quality", particularly regarding Type 53 and Chiang Kai-shek Mausers. Contrary to the belief that these rifles were poorly made, examples of these rifles in good condition demonstrate that they were originally made to excellent quality standards, particularly the Type 53s. Rather, the poor condition of many of these rifles stems from having seen several decades of hard service, perhaps in the hands of different and opposing armies, often with indifferent or inadequate care and storage. One possible scenario for a Chiang Kai-shek Mauser could have been service in World War II from 1937-1945, the Chinese Civil War from 1946-1949, the Korean War from 1950-1953, then Militia service from 1953-1981. This hypothetical rifle saw active service in a total of three 'hot wars' almost continuously from 1937 – 1953, followed by active Militia service in a 'cold war' until 1981, a total of 44 years of continuous service. A Type

---

[14] For more information about the Militia, see the companion book to the Chinese Military Manual Series: Edwin H. Lowe, *Everyone A Soldier! The Chinese Militia 1958-1984* (Sydney: Edwin H Lowe Publishing, 2015)

[15] 'DocAV', 'How did T-53's get to the USA?' Post 16, 20 September 2008, http://forums.gunboards.com/showthread.php?64419-How-did-T-53-s-get-to-the-USA&p=492896#post492896; 'Doc AV', 'Development, Usage, and Collecting the Chinese T-53 Mosin Nagant' Post 161, 16 February 2012. http://forums.gunboards.com/showthread.php?228896-Development-Usage-and-Collecting-the-Chinese-T-53-Mosin-Nagant&p=2044795#post2044795

[16] 'dstorm1911', 'Chinese Type 53 any good????' Post 27, 08 May 2008, http://www.akfiles.com/forums/showpost.php?s=64a17b6aade724f53755eb1a77816798&p=300612&postcount=27; 'DocAV', Why are so many T-53's beaters? Post 19, 06 July 2008, http://forums.gunboards.com/showthread.php?52524-Why-are-so-many-T-53-s-beaters&p=399639#post399639

38 Arisaka could possibly have served even longer in China, with Japanese forces beginning their occupation of Manchuria from 1931; a total of 50 years until its disposal in 1981.

The idea that the Type 53 is an inferior quality weapon is patently incorrect, as examination of examples of 'good' collector grade rifles reveals the high quality of their manufacture. It must be remembered that the Type 53 was built on Soviet supplied equipment and tooling under the guidance of Soviet technical advisors, in order to modernise the People's Liberation Army in the immediate aftermath of the Korean War. The Type 53 carbines were the pride of China's redeveloped and reorganised arms industry, and were manufactured from 1953-1956 during the First Five Year Plan, the most economically prosperous and politically stable period in China between 1949 and 1979. The poor quality of examples of Type 53s can be attributed to their decades of continuous service beginning in 1953 with the PLA, before transferring to the Militia beginning in 1958 and seeing hard continuous service until 1981. A common fallacy or fantasy amongst collectors is that the condition of damaged and worn Type 53 carbines is that it is due to use and abuse "in the jungles of Vietnam". The reality is that the vast majority of Type 53 carbines are former PLA and Militia weapons exported from China in the 1980s. Recently, a large number of Type 53s were exported from Albania in 2011, entering Western markets around 2012, leaving only a small percentage of Type 53s which are actually Vietnam 'bring backs'.[17]

The most common and still prominent misconception amongst collectors relates to the cleaning rod of the Type 53. Some early collectors' examples appeared with a cleaning rod very different from the standard cleaning rod of other Mosin Nagant carbines. Authoritative sources amongst Mosin Nagant collectors reported that the Type 53 carbine had a 17in (43.18cm) cleaning rod with a smooth or plain head, rather than the typical 17.5in (44.45cm) Soviet carbine cleaning rods with a knurled head.[18] In the early days of collecting the Type 53, these sources noted that the so called 17in 'Chinese rod' resembled the SKS cleaning rod and questioned the originality of the 'Chinese rod' versus the 17.5in 'Soviet rod'; whether the 'Soviet rod' found on Type 53s were a replacement of the 'Chinese rod'; and whether the replacement had been made in China or by US importers.[19] Given that the Type 53s were manufactured on tooling and equipment from the Soviet Union, it is almost impossible to consider that Type 53s were not originally manufactured with anything other than the standard 17.5in knurled head carbine cleaning rods. The simplest explanation is that the 17in Type 56 (SKS) cleaning rods were convenient replacements for lost or damaged 17.5in Type 53 cleaning rods, made during the long years of Militia service from PLA spare parts. The fact that many Chinese sourced Type 53s are missing cleaning rods supports the explanation that they have simply been lost. The examples of Type 53s in European and Australian collections since the 1980s which had been directly imported from China, supports the idea that the 17.5in rods could not have been replacements made by US importers. Similarly, the importation to the US of Type 53s from Albania in recent years, further supports the idea the 17.5in rods are original to the Type 53.

Collectors often refer to the wood used in Type 53 stocks as "CHU- wood", most likely repeating a typographical error from a source, which for many years, was the definitive source of information about the Type 53.[20] "CHU-wood" was probably intended to refer to what is more correctly called *qiu* wood in Chinese, *Catalpa bungei* or more commonly, Manchurian Catalpa. Qiu is a light wood, especially resilient to moisture and thus an ideal material for rifle stocks, given the wide geographic and climatic conditions found in China. Qiu wood is found on the wooden furniture of Chinese weapons such as the Type 56 (SKS,) Type 63, Type 56

---

[17] 'LAJMI', *Pushkët kineze, hetime në SHBA* ('Chinese rifles, investigation in the USA'), 23 December 2012, http://gazeta-lajmi.info/pushket-kineze-hetime-ne-shba/; United Nations, The Global Reported Arms Trade, The UN Register of Conventional Arms, National Reports: Albania 2011. http://www.un-register.org/SmallArms/CountrySummaryId.aspx?CoI=2

[18] See Terence W. Lapin, *The Mosin-Nagant Rifle*, 4th Revised and Expanded Edition, (Tustin: North Cape Publications 2007), pp75-77; 'Mosin Nagant Cleaning Rods', http://www.62x54r.net/MosinID/MosinFeatures09.htm; Brent Snodgrass, 'The Chinese Type 53 Mosin Nagant Carbine', http://www.mosinnagant.net/global%20mosin%20nagants/Chinese-T53Carbine.asp

[19] Ibid.

[20] Snodgrass, 'The Chinese Type 53 Mosin Nagant Carbine'

(AK-47) etc with a shellac finish. 'Blonde' coloured when raw, qiu wood stains a 'bright orange' or 'blood orange' with age, presumably subject to moisture or oil content.

A final misconception relates to the storage of Type 53s and other Militia weapons prior to their surplus sale and export. The most common collector's story is that they were stored in 'caves', while other collectors claim that they had been 'buried and dug up' prior to export from China.[21] Taking the 'burial' story as possibly apocryphal, or perhaps referring to isolated cases of 'destructive disposal', we may address the 'cave' story. The idea of weapons being 'stored in caves' conjures the image of pokey, dark and damp caves akin to a Vietcong tunnel complex, with weapons simply thrown in. Reality however, was very different. In the 1964, facing the prospect of a nuclear war with the US and the USSR, China embarked on the 'Third Line' or 'Third Front' project, the construction and relocation of large numbers of strategic defence industries to the remote regions of south western and central China. Faced with the disparities and deficiencies in technology compared to the US and USSR, the Chinese used their experiences in World War II to devise the defence doctrine of People's War. Under the People's War doctrine, the Chinese were prepared to sacrifice the developed industrial north east and coastal regions to American or Soviet nuclear strike or invasion, while the Chinese forces retreated into the vast interior to fight a protracted guerrilla war. In keeping with the experience of World War II and the strategic relocation of China's war industries to the south west, the Chinese constructed new infrastructure and relocated as much as a quarter of its defence industries to the remote 'third line'.[22] A part of these defence installations were vast caverns excavated from the mountains of the south western provinces and entire factories, research institutes and storage areas were located in such caverns. It is more than likely that the reports of the 'cave warehouses' refers to these third line storage facilities.

The long, continuous and active service lives of these Militia rifles over some three to five decades, has left an almost unique but barely appreciated field of interest for collectors of military rifles, accessories and accoutrements. The variety of conditions of Militia rifles, ranging from 'unissued' to 'scrap' is a source of some frustration, particularly for collectors seeking the ideal example of a 'shooter' in good cosmetic condition with matching serial numbers. The reality is however, that the conditions in which collectors find these former Militia rifles, as with all surplus military rifles and their accessories, tell a story about the history of not only the rifles themselves, but also the nations, the forces and the people which carried them. In the case of the rifles of the Militia, that history is a long and complex story of one of the most turbulent and troubled times in all of China's long history. It is an idea that collectors have only recently started to appreciate - and with that appreciation of history, has come the appreciation of those battered and worn Militia rifles as they are, for what they are.

M1903 Springfield with 民兵 Minbing (Militia) branded butt.  Photo: Dan Pickle

---

[21] 'dstorm1911', 'Chinese Type 53 any good????' Post 27, 08 May 2008, http://www.akfiles.com/forums/showpost.php?s=64a17b6aade724f53755eb1a77816798&p=300612&postcount=27; 'tuco', 'Why are so many T-53's beaters?' Post 17 06 July 2008, http://forums.gunboards.com/showthread.php?52524-Why-are-so-many-T-53-s-beaters&p=399622#post399622
[22] Melvin Gurtov and Byong-Moo Hwang, *China's Security: The New Roles of the Military*, (Boulder: Lynne Rienner, 1998), pp148-149

# Militia Markings

Type 38 Arisaka 7.62x39mm conversion. Right side of butt, branded. Photo: Roger Finzel

Chiang Kai-shek Mauser. Right side of butt, branded.

M1903 Springfield. Right side of butt, branded. Photo: Dan Pickle

民兵 *Minbing* Militia.

Type 88 Hanyang, right side of butt, hand carved and branded. Photo. Roger Finzel

来齐堂 *Lai Qitang* 河南 *Henan* 民兵 *Minbing*

Lai Qitang (personal name). Henan Militia

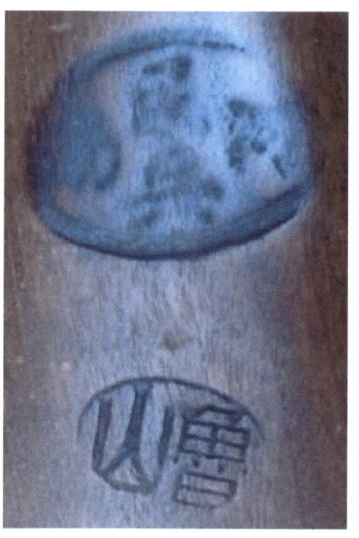

Type 88 Hanyang, right side of butt, two brands. Photo: kalash2

河南 *Henan* 民兵 *Minbing* 鲁山 *Lushan*

Henan Militia. Lushan (county)

This particular oval Militia brand has been the subject of some conjecture amongst collectors. Historically, horizontally written Chinese characters could be read both right to left, and left to right, so that the horizontal characters could be read as either *Henan* (province) or *Nanhe* (several counties). Adding to the uncertainty was that although effective Militia command authority in the PLA began with the provincial Military District, Militia units were organised at the city or county level (division or regiment) and downwards. Although left to right writing gained popularity with the influence of European languages throughout the first half of the 20th century, it was not mandated as standard in the People's Republic of China until 1955. Consequently it has been unclear as to how this brand is to be read, and therefore the identity of the command which issued these rifles has also been unclear. However, the *Lushan* (county) brand on the Type 88 Hanyang on the right, indicates the right to left reading direction in use at the time, and therefore confirms the reading of *Henan Minbing*. Rifles carrying this brand can be identified as having been issued by the Henan Military District.

# People's Armed Forces Department Marking

Gewehr 88 (Loewe), right side of butt, branded.
Photo: Ron Swanson

贵州黄平縣 人民武裝

*Guizhou Huangping Xian. Renmin Wuzhuang*

Guizhou (province) Huangping County.
People's Armed Forces (Department)

The People's Armed Forces Department (*Renmin Wuzhuangbu*) is a dual civil-military organisation of command and control of the Militia. It is both a part of the local civil government and a part of the military chain of command. A PAFD is organised at the each tier of civil government from the city or county levels downwards. Staffed by both People's Liberation Army personnel and local civilian government Communist Party officials, it acts as the permanent cadre staff for local command, control, and training of the Militia. The People's Armed Forces Department also organises and delivers the compulsory annual military training of high school and university students (Junior ROTC and ROTC like programmes) at educational institutions in the local government area. Similarly the PAFD acts as the local PLA recruitment office, manages personnel administration of local PLA reservists, manages the military service registration of local 18 year old males, and acts as the local veteran's affairs office.

This Gewehr 88 is therefore not a Militia weapon as such, but was issued to the People's Armed Forces Department for use in Militia training and instruction by the permanent cadre staff. The Type 88 and other WWII era weapons became progressively obsolete as the ammunition supply of these calibres was exhausted, as early as the late 1950s. They were downgraded to instructional and drill purposes as they were replaced by modern PLA standard weapons. This process included all of the rifles described in this manual, with the Type 53 being the last type remaining in service, by virtue of its use of PLA standard calibre ammunition.

Gaoya Commune, Hui county, Henan province 1974. Photo: Wang Shuzhou

An instructor from the commune People's Armed Forces Department with an obsolete Type 88 Hanyang, while Militia personnel are armed with Type 56 (SKS) carbines. So important was the Militia to the People's War defence doctrine, that PLA personnel were posted as permanent cadre staff to PAFDs at the city or county (division or regiment) level. At the city district or commune (battalion) levels, PAFD personnel were uniformed civilian (ex-PLA) Communist Party cadres (*zhuanwu ganbu*) from the local government. The appearance of the instructor suggests that he is a not a PLA regular, but a CPC cadre from the commune PAFD.

# Public Security Bureau and Militia Markings

Type 53 carbine, left side of butt, branded.
Photo: Edwin H. Lowe

高要县政法公安部 (illegible) 0101 号

*Gaoyao Xian Zhengfa Gonganbu ? 0101 hao*

Gaoyao County Public Security Bureau, ? No. 0101.

The *Zhengfa Gonganbu* is literally translated as 'political and legal public security bureau', which is a good description of the function of this organisation. It existed from the mid-1950s to the early 1960s as an organ of local civil government, with joint responsibility to both the local government and the local Communist Party committee. Its role was that of law enforcement, with a particular emphasis on political security and political control, during the early years of Communist government. This period saw the establishment of new social and political structures throughout China, such as the rural communes and the urban industrial work units. The Public Security Bureau was responsible for political security, such as safe guarding against 'counter-revolutionary activity' and keeping watch on class enemies in the form of former landlords, capitalists, rich peasants and Nationalist Party officials (ie opposition to, and opponents of Communist Party policies). These duties were coordinated by the Public Security Bureau as a local government department, but were typically conducted by the Militia.

Type 53 carbine, right side of butt, painted.
Photo: Edwin H. Lowe

凸庆化肥厂白 *Zhaoqing huafeichang bai*

The reverse side of the Gaoyao County Public Security Bureau Type 53 carbine features this curious hand written text. The first character is 凸 *(zhao)*. Together with the next character 庆 *(qing)*, they can be read as *zhaoqing*, a homonym of Zhaoqing （肇庆）, a town of Gaoyao county, Guangdong province, at the time of the existence of the *Zhengfa Gonganbu*. The use of this homonym could be an 'abbreviation' of 肇庆 Zhaoqing. The following characters, 化肥厂 *(huafeichang)* are 'fertiliser factory'. The final character appears to be 白 *(bai)* 'white', and it appears that there was once a further character, now illegible. Baitu township （白土）is a part of the Zhaoqing - Gaoyao county area, so it is possible that the 白 character refers to Baitu township. It is conceivable that this handwritten text stands for *Zhaoqing huafeichang Baitu* (Zhaoqing Fertiliser Factory Baitu). If this were the case, it would indicate that this particular Type 53 carbine was issued to the Gaoyao County Public Security Bureau and possibly later transferred to a Militia unit established in the fertiliser factory in the same county. If so, it would be an identification of service issue down to the Militia small unit level, possibly company, platoon or squad.

# Political Slogans

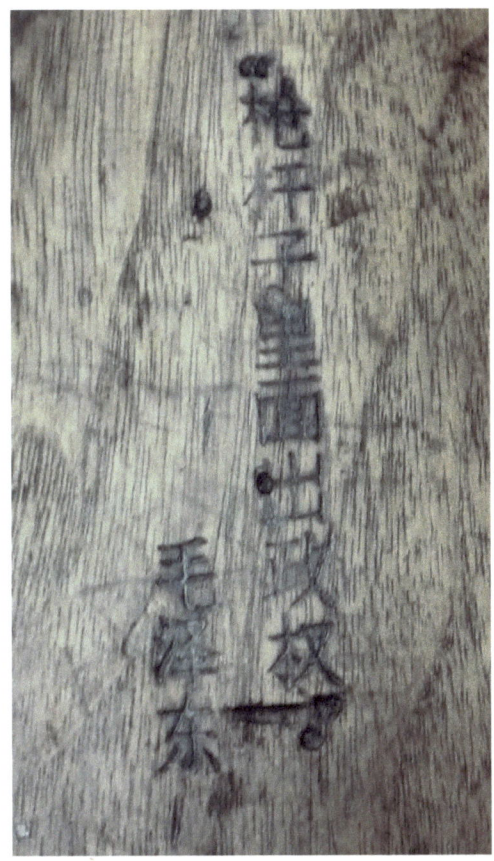

Type 53 carbine, right side of butt, branded.
Photo: CR Holt

"枪杆子里面出政权"。 毛泽东

"*Qiang ganzi limian chu zhengquan*". Mao Zedong

"Political power grows from the barrel of a gun". Mao Zedong

This famous quote from Mao Zedong's article 'Problems of War and Strategy' (November 6, 1938) was reproduced in *Quotations From Chairman Mao Zedong*, more famously known as 'The Little Red Book'. Essentially required reading during the Cultural Revolution (1966–1976), this dates the marking to that period and its Mao cult. It is more likely to be from the earlier period (1966-1969) when the cult reached its apogee.

Militia studying the 'Little Red Book' with Type 53 carbines and Type 50 SMGs, 1970.
Photo: *China Reconstructs*

Type 53 carbine, right side of butt, painted.
Photo: Danny Nichols

忠 *zhong* Loyalty.

This marking is likely to date to the Cultural Revolution, when such 'decorations' were common.

# Damage to Militia Rifles

Firearms industry traders who participated in the process of exporting the Militia's surplus bolt action rifles from China in the 1980s, reported on the conditions of their storage and transport. Observers noted that rifles were bundled in groups of 10, then tied with bailing wire and 'tourniqueted' with a stick, with 20 such bundles wired together onto a pallet. These conditions left the rifles, particularly the soft qiu wood stocks of Type 53s, literally with 'lasting impressions'. In addition to the typical 'dings and dents' to stocks during use and storage, Chinese exports of surplus Type 53s feature some characteristic damage to wooden stocks. The gouge on the photo on the right was left by the highly tensioned bailing wire used to tie and bundle rifles together onto pallets. Another feature of damage to Chinese exports is the curious serrated oval imprints. Closer inspection of this example's fore-end reveals that it is the exact impression of the grip of the backsight slide catch of the Type 53 carbine. This is revealing of the amount of tension used to bundle the rifles together with bailing wire and the weight of stacked bundles on their pallets.

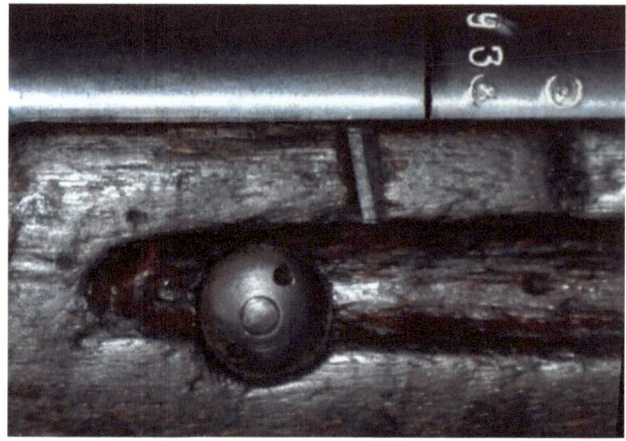

Chinese exported Type 53s (above) and Chiang Kai-shek Mausers (below) exposed to rain in the open, with stocks in a shipping container at an Arizona firearms dealer, 2007. Photos: Ban Pingcu

Damage to Type 53 stock. Photos: Edwin H. Lowe

# Type 53 Carbine Sling

Type 53 sling, unmarked.  Photo: Edwin H. Lowe

The Type 53 carbine was originally issued to the PLA with a typical Mosin Nagant pattern webbing sling. However the Chinese Type 53 sling differs from the Soviet M44 sling in two details. The Chinese sling has sling loops which are secured by cord, rather than the Soviet style buckles. Additionally, the Chinese slings are 3cm (1.18in) wide, compared to the Soviet slings which are 3.5cm (1.38in) wide.[23]

The simplest explanation as to why the sling loops are cord tied rather than buckled, would be that in 1953, China was only just beginning to recover from the devastation of its industry and economy, after over 40 years of continuous revolutions, foreign invasion and civil wars. A cord tie would simply have been cheaper and more easily replaceable than a buckle, when damaged or lost.

The reason for the difference in sling widths is not immediately obvious, until a Type 53 sling is compared to the generic webbing sling used by the Militia for its other bolt action rifles. The 'generic' sling is 3cm wide, in order to fit the sling swivels of a variety of rifles such as Arisakas, Mausers and Springfields. Both types of slings show the same factory marking, 囦$\frac{11}{3}$. Therefore it is reasonable to assume that it was simpler and more economical to use the same width material for both patterns of slings, as well as for the slings of the Type 56 (SKS) carbine and Type 56 (AK-47) assault rifle.

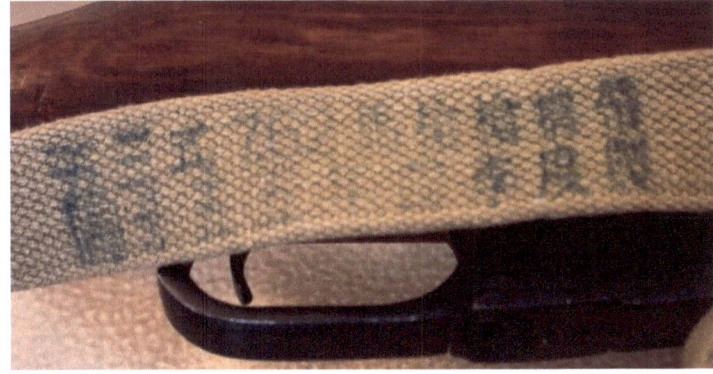

Type 53 sling, ink stamped.  Photo: Bob Hanes

| | | |
|---|---|---|
| 五三式 7.62 馬槍背帶 | *Wusan shi 7.62 maqiang beidai* | Type 53 7.62 Carbine Sling |
| 囦$\frac{11}{3}$ 五五年四季度製 | 囦$\frac{11}{3}$ *Wuwu nian si jidu zhi* | 囦$\frac{11}{3}$ Manufactured 4th Quarter 1955 |

---

[23] As with Type 56 SKS & AK slings.

# Militia Generic Rifle Sling

The generic webbing sling was issued as a replacement for the original slings of a variety of bolt action rifles in Militia service. The sling's 3cm width allowed it to fit through all rifle swivels, and it was secured at the leather reinforced end by a cord tie. Different ink stamps designating the rifle type for the sling are known, although the patterns of the slings are the same. The generic sling was often found on Type 53 carbines through the use of wire loop swivels.

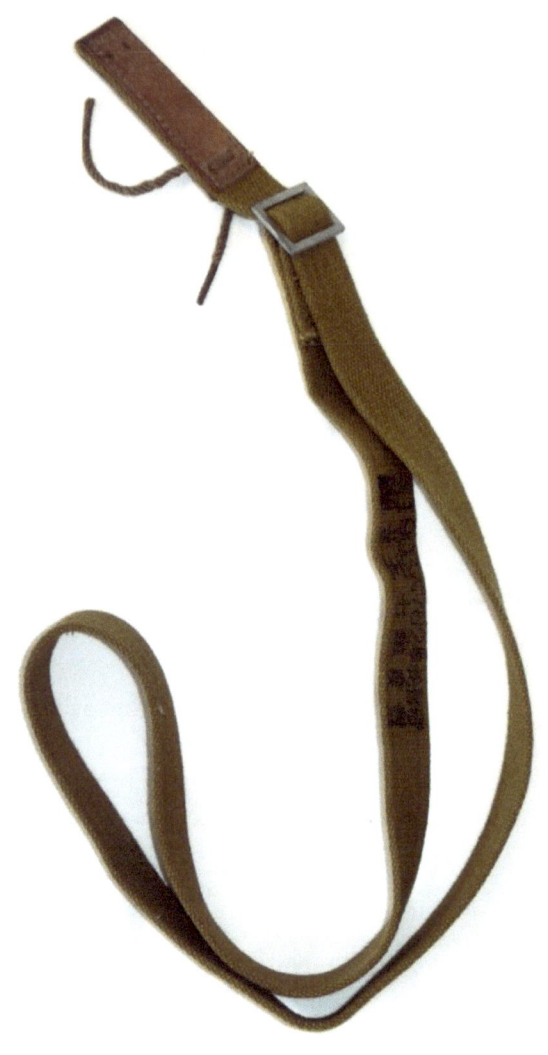

Type 53 with wire loop. Photo: Richard Babb

Type 53 with wire loop and the remains of a US M1 sling. Photo: John Jett

Generic sling, ink stamped for M1903 Springfield.
Photo: Edwin H. Lowe

美 1903 式 762 步枪背带    囶 $\frac{11}{3}$ 五七年度制

*Mei 1903 shi 762 buqiang beidai*    囶 $\frac{11}{3}$ *Wuqi niandu zhi*

US M1903 7.62 Rifle Sling    囶 $\frac{11}{3}$ Manufactured 1957

Photo: Edwin H. Lowe

三八式步枪背带

*Sanba shi buqiang beidai*

Type 38 Rifle Sling

Photos: Teri Jane Bryant  www.nambuworld.com

# Variations on Militia Slings

Despite the standard patterns of slings for Militia weapons, the Militia's decentralised structure and logistics system meant that in reality, a variety of slings were used on different weapons. The following examples c.1970 will be of interest to what collectors constitute as the 'correct' combination of slings and weapons.

This Militia squad in southern China reveals the typical mixture of slings in use. The first Militiawoman carries a Type 50 SMG featuring a Type 56 SKS/AK rifle sling. The rest of the squad carry Type 53 carbines. The second and third Militiawomen carry carbines with a wire loop and generic sling, while the fifth Militiawoman carries a carbine with a Type 50/54 SMG sling. The Type 56 rifle sling appears to have become the standard replacement SMG sling by the late 1960s.

This Militiawoman on a South China Sea island carries a Type 53 with a Type 50/54 SMG sling.

This Militiaman in north eastern China carries a Type 53 carbine with wire loop and generic sling.

This Militia squad in north western China carry Type 54 SMGs with Type 56 rifle slings.

# Type 53 Carbine Cleaning Kits

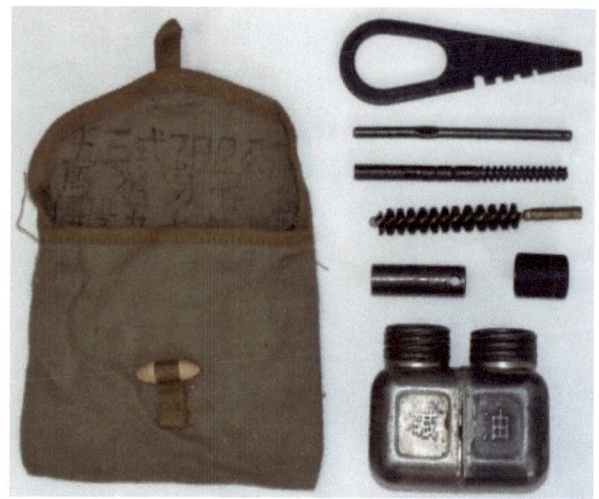

An original pattern Type 53 carbine cleaning kit, as described in the 1973 Militia training manual. In this example, only the 'accessories pouch' and the oil bottle are identifiable as original Chinese manufactured items. The cleaning tools are either unmarked or Soviet in origin. The pouch features a wooden toggle and loop fastener, and a belt loop on the reverse side. The nature of cleaning kits with small, easy-to-lose items means that few original cleaning kits, complete or partial, survive with collectors after the continuous service of the Type 53 carbine from 1953 to 1981.

Original pattern Type 53 carbine accessories pouch and cleaning kit.   Photo: Edwin H. Lowe

| | | | |
|---|---|---|---|
| 五三式 7.62 公厘 | 馬槍另件袋 | 囥 $\frac{11}{3}$ 五七年度製 | Double cell oil bottle |
| *Wusan shi* 7.62 *gongli* | *maqiang lingjiandai* | 囥 $\frac{11}{3}$ *Wuqi niandu zhi* | Left:   碱 *jian*  alkali |
| Type 53 7.62mm | Carbine Accessories Pouch | 囥 $\frac{11}{3}$ Manufactured 1957 | Right:  油 *you*  oil |

Late pattern Type 53 carbine accessories pouch and cleaning kits.
Photos: Craig Kratzer (left, centre) and James E. Wilen Jr (right)

Characteristic of these late pattern cleaning kits are the use of plastic Type 56 oil bottles and a generic screwdriver tool with a single notch wrench. The kit in the centre features a slotted head cleaning jag, more common with later Type 56 rifles. These kits reveal the logistical benefits of interchangeability through standard design principles and the use of common tools in PLA rifle cleaning kits. All of the tools could be used interchangeably with the standard PLA rifles; the Type 53, Type 56 (SKS), Type 56 (AK-47) and Type 63. This interchangeability is exemplified in the fitting of Type 56 (SKS) carbine cleaning rods (17in, small smooth head) on some Type 53 carbines as replacements for lost original rods (17.5in, knurled head). Although some tools such as the cleaning jag and bore brush are common issue to all standard PLA rifles, the use of the Type 53 cleaning rod collar, handle/drift punch and muzzle guide reveals the probable intended issue of these kits with the Type 53. However, since all of these tools will fit any standard PLA rifle cleaning rod, it is possible that these are generic replacement cleaning kits to suit any standard PLA rifle in Militia service. The kit on the right features a pull through in addition to the cleaning rod tools, which would allow this kit to be used with non-standard Militia weapons such as Mausers and Arisakas, or perhaps a standard rifle missing its cleaning rod.

# Aim Corrector (Universal)

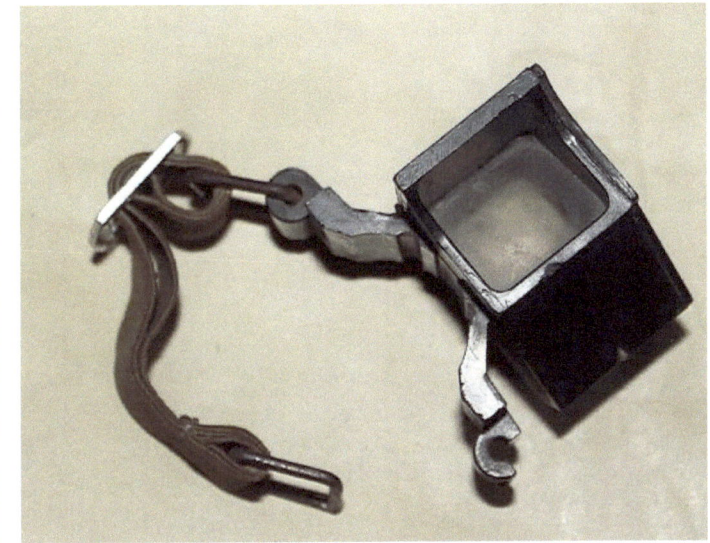

| | | |
|---|---|---|
| 53 式步骑枪 | 53 *shi buqiqiang* | Type 53 Carbine |
| 56 式冲锋枪 | 56 *shi chongfengqiang* | Type 56 Assault Rifle |
| 56 式半自动步枪 | 56 *shi banzidongqiang* | Type 56 Semi-automatic Carbine |
| 63 式自动枪 | 63 *shi zidongqiang* | Type 63 Automatic Rifle |
| 通用瞄准检查镜 | *Tongyong miaozhun jiancha jing* | Universal Aim Corrector |
| 勿用硬物擦拭镜片 | *Wu yong ying wu cashi jingpian* | Do Not Clean Mirror With Hard Objects |

The Aim Corrector (Universal) was designed for use with all standard rifles in the PLA and Militia inventories. This pattern of aim corrector dating to c.1970 with the rifle types indicated, reiterates the standardisation of weapons and ammunition calibres (7.62x54mmR & 7.62x39mm) in the PLA and Militia throughout the 1960s. This was essentially achieved by 1970, with the vast majority of non-standard or 'obsolete' Militia rifles such as Mausers, Arisakas and Springfields etc, relegated to drill and instructional purposes. This pattern of aim corrector is depicted in the 1973 Militia training manual in use with the Type 53 carbine.

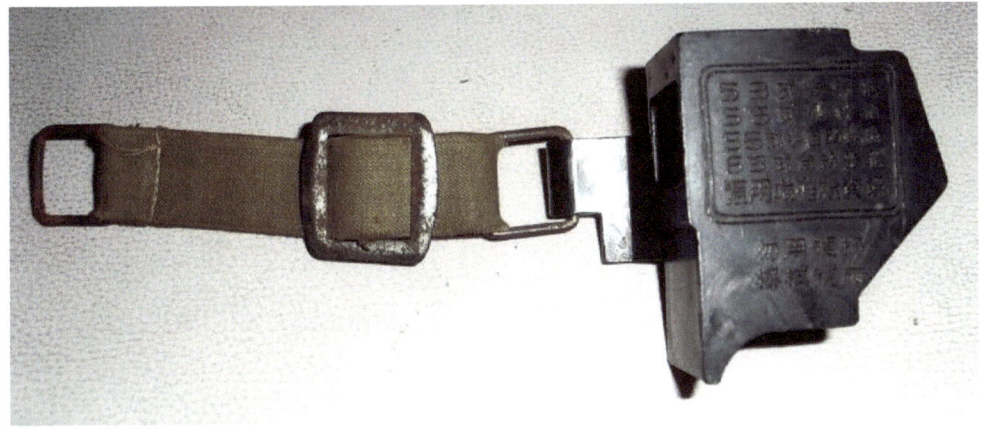

# Type 53 Carbine Ammunition Pouches

The Type 53 carbine 'chest rig' ammunition pouches were standard issue for 'full power cartridge' Militia rifles by the mid-1960s, replacing the generic cotton bandoliers which were standard in the 1950s. The pouches carried 100 rounds with two 5 round chargers in each pouch. There were two patterns of pouches; the early and more common tie fastened pattern, and the later pattern featuring a wooden toggle and loop fastener in the style of the Type 56 (SKS) carbine pouches.    Photos: Edwin H. Lowe

Ammunition Pouches, ink stamped.

㊔ 3/3 步（騎）搶弹袋　　一九五六年一季度製

㊔ 3/3 *bu(qi)qiang dandai*　　*Yijiuwuliu nian yi jidu zhi*

㊔ 3/3 Rifle (Carbine) Ammunition Pouches Manufactured 1st Quarter 1956

Ammunition Pouches, ink stamped.

五三式 7.62 毫米步马枪子弹袋
六０一工厂 1960 年度制

*Wusan shi* 7.62 *haomi bumaqiang zidandai*
*Liu0yi gongchang* 1960 *niandu zhi*

Type 53 7.62mm Carbine Ammunition Pouches Factory 601 Manufactured 1960

Both these sets of ink stamped ammunition pouches are of the same early tie fastened pattern. Note that the 1956 set is marked as 'Rifle (Carbine)' while the 1960 set is specifically marked as 'Type 53 carbine'.

# Submachine Gun Magazine Pouches

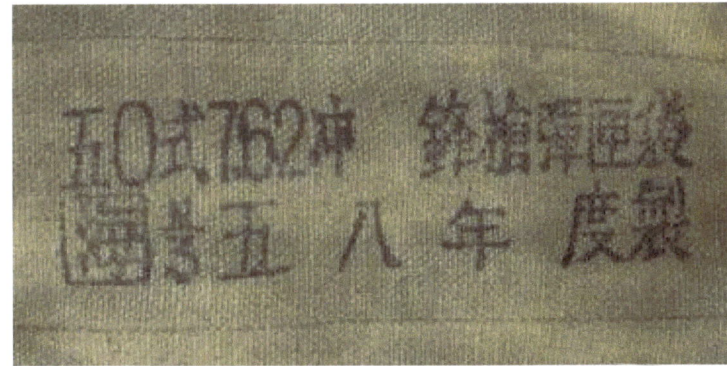

Type 50 SMG Magazine Pouches, ink stamped.

五０式 7.62 冲鋒槍彈匣袋
㠙 11/3 五八年度製

*Wu0 shi 7.62 chongfengqiang danxiadai*
㠙 11/3 *Wuba niandu zhi*

Type 50 7.62 Submachine Gun Magazine Pouches
㠙 11/3 Manufactured 1958

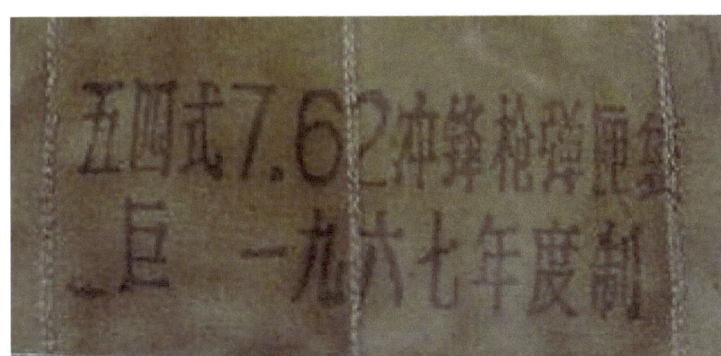

Type 54 SMG Magazine Pouches, ink stamped.

五四式 7.62 冲锋枪弹匣袋
巨 一九六七年度制

*Wusi shi 7.62 chongfengqiang danxiadai*
巨 *Yijiuliuqi niandu zhi*

Type 54 7.62 Submachine Gun Magazine Pouches
巨 Manufactured 1967

Type 50 pouches (left) and Type 54 pouches (right).

The Type 54 pouches include an additional two short length accessories pouches for the cleaning kit and oil bottle, arranged vertically. This was a requirement for the folding stock Type 54 SMG, which had no butt stock cleaning kit storage facility, unlike the Type 50 SMG.

There were two general patterns of magazine pouches; tie fastened, and wooden toggle and loop fastened. Both these tie fastened pouches (left) were made by factory 巨 in 1967, although factory 㠙 11/3 was producing toggle and loop fastened pouches as early as 1956 (top and bottom photos).

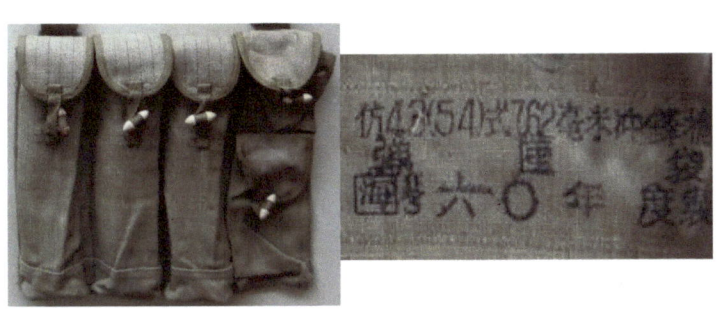

Type 54 pouches, ink stamped. Factory 㠙 11/3 1960

仿 43（54）式 7.62 毫米冲锋枪，弹匣袋

Type 54 SMG receivers were marked 仿 43 式 (copy of the Type 43). This ink stamp refers to that designation. An earlier (1956) variation of ink stamp from the same factory used the designation 四三式 (Type 43). Photos: Xue Rencan

# Ammunition

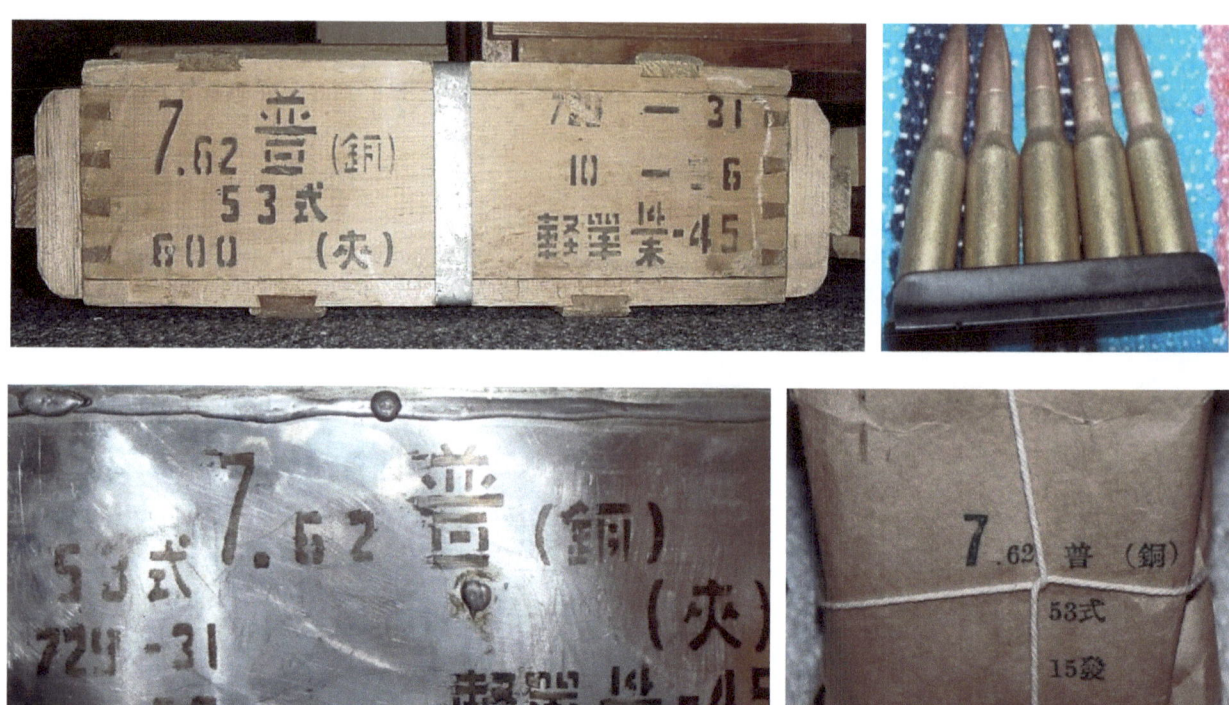

Type 53 7.62x54mmR Lead Core Light Ball, Factory 31, 1956. Photo: Ralph Simmons (upper left), iwannasee

The Type 53 Light Ball cartridge features a steel jacket, lead core, flat base bullet (Soviet Type L) with a brass case. The information layout on the crate and can is in the Soviet pattern and indicates ball cartridges ( 普 *pu* ) with brass cases ( 銅 *tong* ) in chargers ( 夾 *jia* ). Lot–factory (729-31), manufacture date (10–56), propellant type ( 輕單 ) and propellant lot/year-factory ( $\frac{14}{未}$ – 45 ) are also indicated. 普 is an abbreviation of 普通弹 *(putongdan)* meaning 'ball cartridge'. Lead core light ball was designated 7.62 普 (ball) and labelled in the Soviet pattern in 1956 only. From 1952–1955, the cartridge was designated 7.62 輕彈 (light ball) and apart from 夾 (chargers), no other information appeared on ammunition labels. Production of lead core light ball ceased in 1956, as steel core light ball (Soviet Type LPS) designated 7.62 钢轻, became the new production ball cartridge. The propellant year of manufacture is indicated by 未 *(wei)*, which suggests 'no data' or 'not recorded'. The propellant type 輕單 *(qing dan)* is marked on the labels of all Type 53 7.62x54mmR cartridges from 1956 to 1965. 輕單 appears to be an acronym, where 輕 (light) indicates that it is a propellant for use in 'small (light) arms' ( 轻武器 ) ammunition, and 單 (single) indicates that it is a single tube ( 单孔 ) propellant grain.[24] On ammunition labels of post-1965 Type 53 steel core light ball, redesignated 7.62 普, this specification is clearly indicated in the new propellant designation 3/1 樟.[25] Similarly, the equivalent Soviet Type L and LPS light ball cartridges (as with all Soviet 7.62x54mmR cartridges) are labelled in Cyrillic with the propellant type ВТ. This is believed to be an acronym for винтовочный трубчатый *(vintovochnyy trubchatyy)* meaning 'rifle tubular', where 'rifle' indicates 'rifle propellant' and 'tubular' indicates the tubular propellant grain.[26] The correlation of 輕單 with the propellant designations of both Soviet and later Chinese 7.62x54mmR ammunition, suggests that 輕單 is an acronym for 'small arms ( 轻武器 ) single tube ( 单孔 ) propellant'. Each wooden crate contains two cans of 300 rounds, each with 20 paper packets of 15 rounds ( 發 *fa* ) in three chargers.

---

[24] Type 53 cartridges are used in a range of small & light arms: Type 53 Carbine, Type 53 (DPM) LMG & Type 53 (Goryunov) MMG.
[25] See also Type 53 7.62x54mmR Steel Core Light Ball Factory 71, p79; and Factory 81, p80.
[26] Personal communication, Tom Kulik; http://kk-combat.ucoz.ru/proekt1/Htm_boot/m_762v.html

Type 53 7.62x54mmR Steel Core Light Ball, Factory 71, 1968.
Photos: Stevo, www.milsurps.com

The Type 53 Steel Core Light Ball cartridge features a steel jacket, steel core, boat tail bullet (Soviet Type LPS) with a steel case. In 1956, steel core light ball became the standard production light ball cartridge, as lead core light ball (Soviet Type L) designated 7.62 普 ceased production. The steel core light ball cartridge was produced from 1956-1965 under the designation 7.62 钢轻, with a silver bullet tip colour code.[27] In 1965, steel core light ball was redesignated 7.62 普 without any bullet colour coding.[28] The information on the can is in the Soviet pattern and indicates ball cartridges ( 普 *pu* ) with steel cases ( 铁 *tie* ). Lot–factory (0007-71), manufacture date (3-68), propellant type ( 3/1 樟 ) and propellant lot/year-factory ( $\frac{15}{67}$ 25 ) are also indicated. The propellant type is coded 3/1 樟, where '1' indicates that it is a single tube propellant grain, and '3' indicates the thickness of the wall of the tubular propellant grain in multiples of 0.1mm, ie a thickness of 0.3mm. 樟 *(zhang)* meaning 'camphor', refers to the camphor deterrent coating of the propellant. Each wooden crate contains two cans of 440 rounds, each with 22 paper packets of 20 rounds. The crates are externally marked in Albanian, replicating the original Chinese cartridge and manufacturing data.

---

[27] See also Type 53 7.62x54mmR Steel Core Light Ball Factory 81, p80
[28] By 1973 when the Militia manual in this book was published, the supply of ball ammunition included lead core light ball (L) and steel core light ball (LPS) both designated 7.62 普, along with 'silver tip' steel core light ball (LPS) designated 7.62 钢轻. This logistic situation is reflected in the table of Type 53 ammunition types, in Appendix 6.A of the manual.

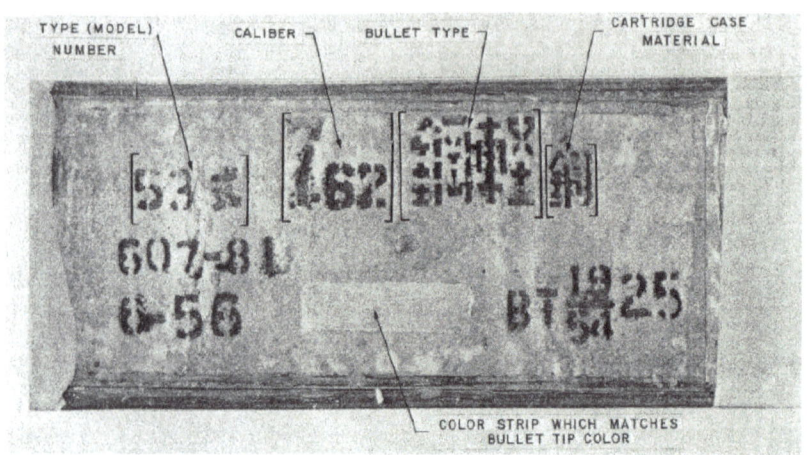 

Type 53 7.62x54mmR Steel Core Light Ball, Factory 81, 1956. Photos: Department of the Army [29]

The Type 53 Steel Core Light Ball cartridge (Soviet Type LPS) was designated 7.62 钢轻 from 1956-1965, during which time the designated ball cartridge 7.62 普, was the lead core light ball (Soviet Type L). 钢轻 (*gang qing*) is an acronym of 钢心轻弹 (*gangxin qingdan*), where 钢 indicates 'steel core' (钢心) and 轻 indicates 'light ball' (轻弹). The information on the can indicates steel core light ball (鋼輕) cartridges with brass cases (銅 *tong*). The bullet tips and colour strip on the can and crate are colour coded silver, indicating steel core light ball cartridges. Of particular note is the ammunition crate which gives the propellant type as 輕單, whereas the ammunition can gives the propellant type in Cyrillic as BT, as found on Soviet 7.62x54mmR ammunition cans. This further supports the suggestion that Soviet BT, Chinese 輕單 and Chinese 3/1 樟 are all designations of the same propellant type, and that BT and 輕單 are both acronyms, essentially conveying the same meaning.

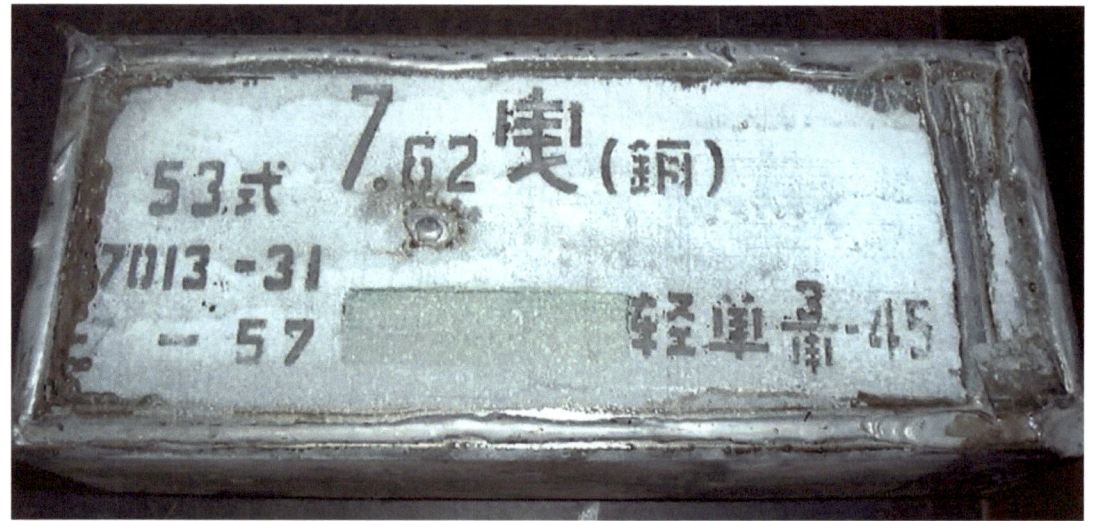

Type 53 7.62x54mmR Tracer, Factory 31, 1957. Photo: Howard A. Bearse

The Type 53 Tracer cartridge features a steel jacket, boat tail bullet, with a tracer component in the base of the bullet (Soviet Type T-46). The information on the can indicates tracer cartridges (曳 *ye*) with brass cases (铜 *tong*). Lot–factory (7013-31), manufacture date (7-57), propellant type (轻单) and propellant lot/year-factory ($\frac{3}{末}-45$) are also indicated. 曳 is an abbreviation of 曳光弹 (*yeguangdan*), meaning 'tracer cartridge'. The bullet tips and colour strip on the can are colour coded green, indicating tracer cartridges.

---

[29] Headquarters, Department of the Army, *Department of the Army Pamphlet 381-12. Recognition Guide of Ammunition Available To, Or In Use By, The Vietcong*, (Washington D.C.: Department of the Army), May 1966, p32 & p34.

Type 53 7.62x54mmR Armour Piercing Incendiary, Factory 71, 1958.
Photos: Howard A. Bearse

The Type 53 Armour Piercing Incendiary cartridge features a steel jacket, boat tail bullet, with an incendiary component in the tip, followed by a steel core penetrator (Soviet Type B-32). The information on the can indicates Armour Piercing Incendiary cartridges ( 穿燃 *chuan ran* ) with brass cases ( 铜 *tong* ). Lot–factory (8002-71), manufacture date (2-58), propellant type ( 輕单 ) and propellant lot/year-factory ( $\frac{32}{末}$ – 25 ) are also indicated. 穿燃 is an acronym of 穿甲燃烧弹 *(chuanjia ranshaodan)*, where 穿 indicates 'armour piercing' ( 穿甲 ) and 燃 indicates 'incendiary cartridge' ( 燃烧弹 ). The bullet tips and colour strip on the can and crate are colour coded black and red, indicating Armour Piercing Incendiary cartridges.[30] Each wooden crate contains two cans of 440 rounds, each with 22 paper packets of 20 rounds.

---

[30] See Appendix 6.A in the Militia manual for the full range of Type 53 7.62x54mmR ammunition colour codes.

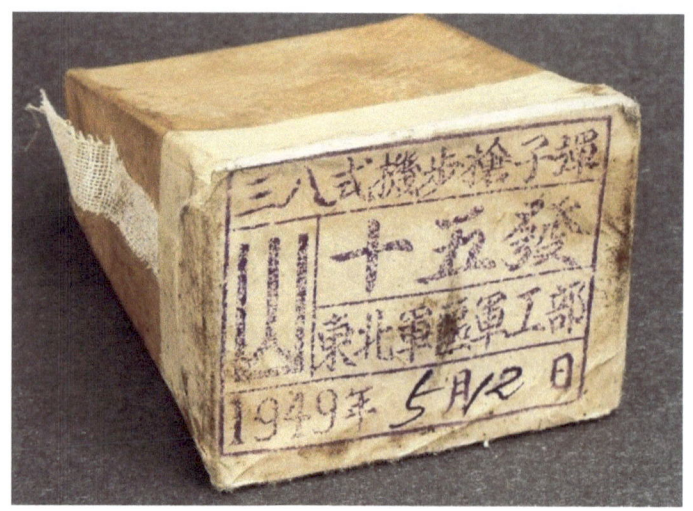 

Chinese 6.5mm Arisaka Ball.

| | | |
|---|---|---|
| 三八式幾步搶子彈 | *Sanba shi jibuqiang zidan* | Type 38 Machine Gun and Rifle Cartridges |
| 山 十五發 | 山 *shiwu fa* | 山 15 Rounds |
| 東北軍區軍工部 | *Dongbei Junqu Jungongbu* | North East Military Region, Bureau of Military Industry |
| 1949年5月12日 | 1949 *nian* 5 *yue* 12 *ri* | 12 May 1949 |

Following the Japanese surrender in 1945, the Mukden (Shenyang) Arsenal in Manchuria which had been under Japanese occupation since 1931, returned to Nationalist control. After the resumption of the Civil War in 1946, it passed into Communist control in 1948. Under Communist control, the arsenal produced 7.92mm Mauser and 6.5mm Arisaka ammunition until around 1952, when production ceased in favour of the new standard PLA calibres (7.62x54mmR and 7.62x25mm). This 6.5mm Arisaka ammunition was produced by the Communists during the Civil War and features the green case sealant typical of Chinese production from 1948-1952. Relatively little of the Mauser and Arisaka ammunition made in China by any government survived to be acquired by collectors, as most of it was used by the PLA during the Civil War and Korean War, with the Militia exhausting the remaining supply during the 1960s.

 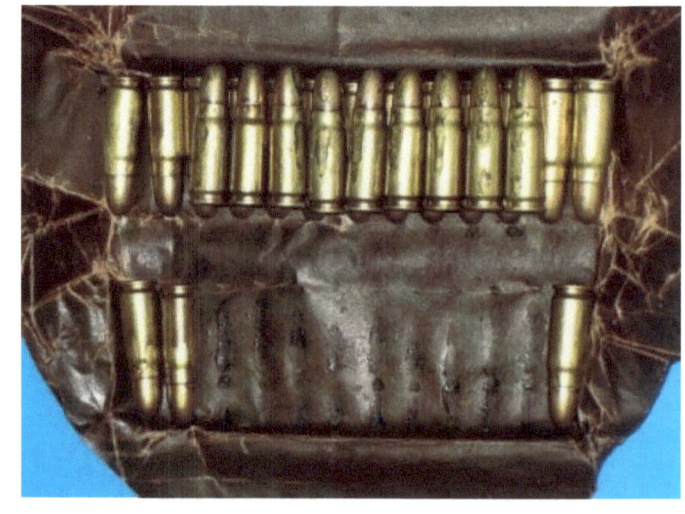

7.62x54mmR Drill Cartridges   7.62x25mm Drill Cartridges

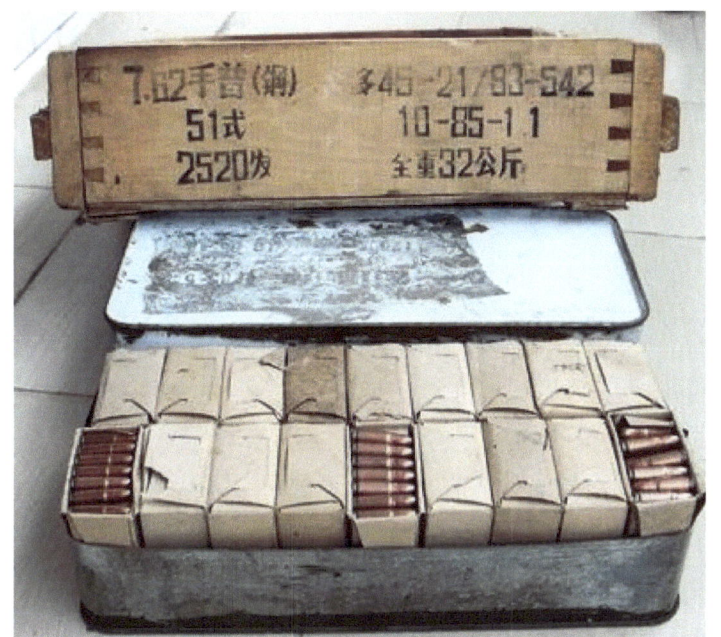

Type 51 7.62x25mm Ball.  Photo: Sun Jirong

The Type 51 Ball cartridge (Soviet Type P) features a steel jacket, lead core bullet with either a brass or steel case. The information on the crate is in the Soviet pattern and indicates pistol ( 手 *shou* ) ball cartridges ( 普 *pu* ) with steel cases ( 钢 *gang* ). Month–year-factory (10-85-11), propellant type (多 45) and propellant lot/year-factory (21/83-542) are indicated, as are 2520 rounds (发) and total weight (全重) of 32kg (公斤). 手 is an abbreviation of 手冲 *(shouchong)*, itself an acronym meaning 'pistol and SMG'. The propellant type 多 45 is presumably the same propellant as the п-45 in Soviet 7.62 P cartridges. п is assumed to stand for пистолет (pistol), following the same convention for the 7.62x54mmR cartridge where the propellant type BT is an acronym of винтовочный трубчатый ('rifle tubular'), indicating its use in rifle cartridges.[31] Maintaining the Soviet convention, the Chinese designation of the 7.62x25mm cartridge is '51 式手普' an acronym for 'Type 51 Pistol and SMG Ball Cartridge'. 多 *(duo)* in the 多 45 propellant type means 'many', and the meaning of this can be understood in the context of the Type 51 Ball cartridge being used in 'many' types of pistols and submachine guns, including the Type 51 and Type 54 (Tokarev), the Type 50 SMG and Type 54 SMG. Each wooden crate contains two cans of 1260 rounds, packed in 18 boxes of 70 rounds.

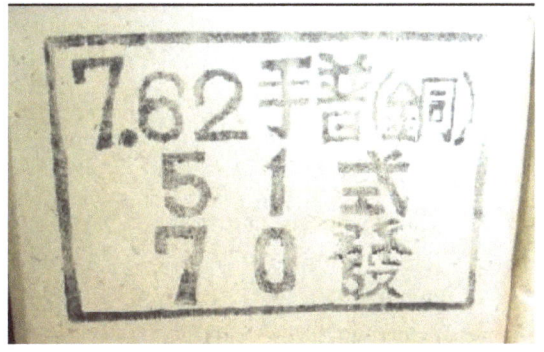

These 70 round ( 發 ) boxes of Type 51 Ball ( 51 式手普 ) indicate (left) pre-1965 brass cartridge cases ( 銅 ), and (right) post-1965 steel cartridge cases ( 钢 ).

7.62x25mm Type 50 SMG Cartridges, Factory 121.
Photo: Dan Zhuqu

At the time of its introduction with the Type 50 SMG, 7.62x25mm ball (left) was labelled 'Type 50 SMG Cartridges' ( 五〇式衝鋒槍彈 ). This was a necessity due to the large variety of calibres and weapons in service prior to standardisation. After standardisation, it was designated 'Type 51 Pistol and SMG Ball' ( 51 式手普 ).

---

[31] Personal communication, Tom Kulik; http://kk-combat.ucoz.ru/proekt1/Htm_boot/m_762v.html

# Militia Identification Badges

From 1958 to c.1998, the Militia did not wear a standard uniform and in most cases, duties ranging from training to patroling and combat were performed in civilian clothing. The only uniform item was a chest badge, typically in the pre-1955 PLA uniform style or an armband, but even the issue and wearing of these appeared to be inconsistent or arbitary in nature. Typically the front of the chest badge indicated the Militia unit, while the back of the badge recorded, name, serial number, age, sub-unit and appointment.[32]

 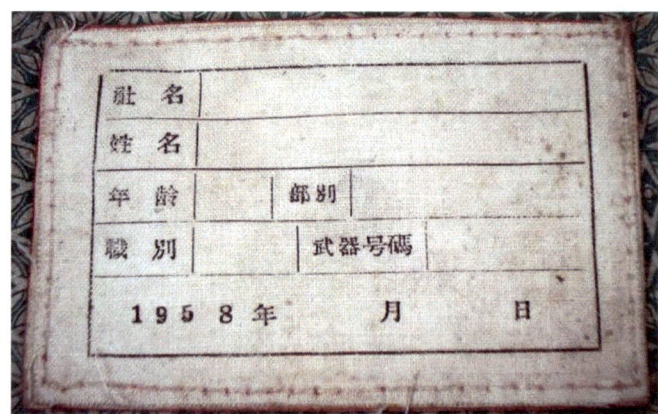

Ordinary Militia 普通民兵, 1958.

In 1958 at the beginning of the Everyone A Soldier movement and the massive expansion of the Militia organsation, personnel were organised into two groups, the Ordinary Militia (*Putong Minbing*) and the Primary Militia (*Jigan Minbing*). Only the Primary Militia personnel, men aged 16-30 and women 16-25 who passed the stringent eligibility requirements of class (eg lower and middle peasant, workers, ex-PLA personnel) and political reliability (eg 'clean' family political backgrounds, Communist Youth League or Communist Party members), received military training. The Primary Militia consisted of an estimated 100 million men and women by 1981. Of these, only the Armed Primary Militia (*Wuzhuang Jigan Minbing*) were issued individual weapons and performed military or public security duties, with an estimated 15 million personnel by 1981.

The Ordinary Militia existed largely as a 'paper organisation' of able bodied men and women (less 'class enemies' of former landlords, rich peasants, capitalists and Nationalist government officials), organised as a part of the general militarisation of Chinese society in 1958. The Ordinary Militia received no military training and acted essentially as an militarised and politically indoctrinated population for the purposes of economic development, numbering an estimated 150 million in 1981. Its military role was limited to acting as a system for a general wartime mobilisation of the entire population under the People's War doctrine.[33]

Given that the Ordinary Militia was a paper organisation and that the Militia's logisitics system was decentralised, this badge can be considered an extremely rare example and it can be assumed that only a few were ever made. The badge shows the symbol of the Militia, the hoe and rifle, which reflects the multi-faceted roles of defence, economic production and social organisation under the Everyone A Soldier movement. The red star and red flag symbols are those of the Communist Party of China, rather than the People's Liberation Army. This reflects the ultimate command of the Militia being under the authority of the CPC rather than the PLA, in accordance with Mao Zedong's maxim of "Political power grows from the barrel of a gun. Our principle is that the Party commands the gun, and the gun must never be allowed to command the Party".[34]

---

[32] eg Platoon Commander, Militiaman etc. There has never been a system of rank in the Militia, even when it has existed in the PLA.
[33] For more comprehensive information about the Militia in this period, see the companion book to the Chinese Military Manual Series, *Everyone A Soldier! The Chinese Militia 1958 – 1984*, by Edwin H. Lowe, (Sydney: Edwin H. Lowe Publishing, 2015)
[34] Mao Zedong, 'Problems of War and Strategy' (06 November 1938), in *Selected Works of Mao Tse-tung*, Vol. II (Peking: Foreign Languages Press 1969), pp. 224-225. See also the Type 53 carbine butt branded with this quotation, p69.

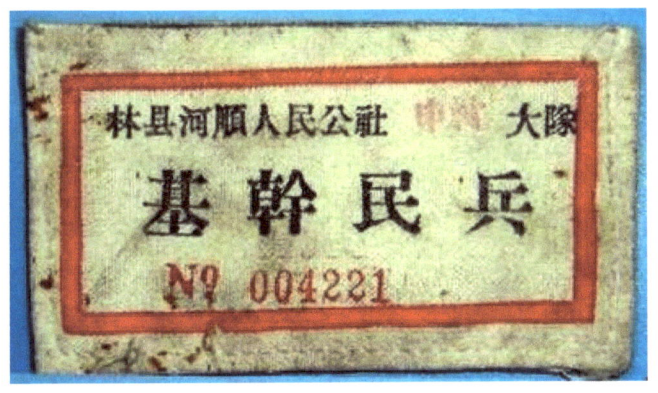 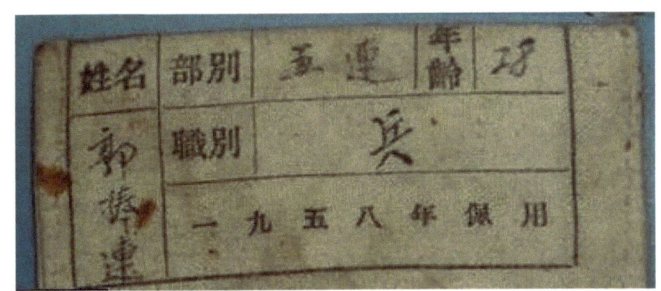

Lin County 林县, Heshun People's Commune 河顺人民公社, Primary Militia 基幹民兵, 1958

This badge contains the complete issue information to Militiaman（兵）Guo Fenglian（郭俸連）, serial number 004221, aged 28. In rural areas, the Militia was organised at the county level into divisions (or regiments) from battalions formed in the communes. In this case, the red ink stamped name (申 ?) of the village(s) within the commune which formed the battalion（大隊）is only partially legible. The battalion sub-unit is No. 5 Company（五連）.

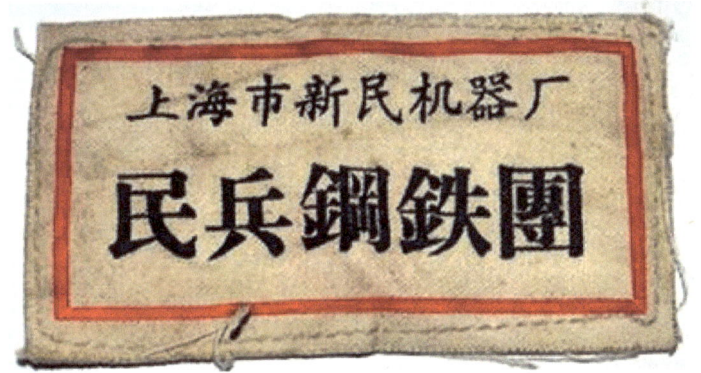 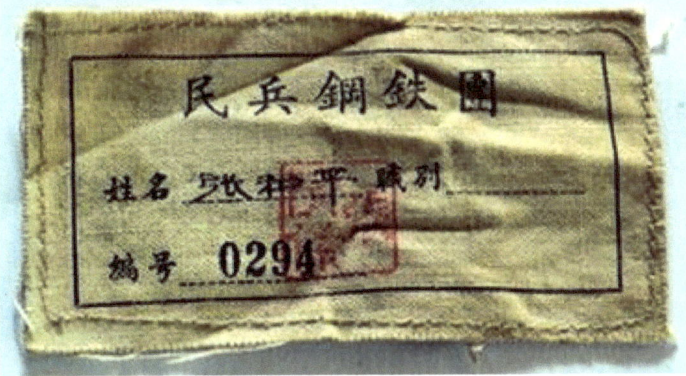

Militia Steel Regiment 民兵鋼鉄團, Shanghai New People's Machinery Factory 上海市新民机器厂, c.1958

In the urban or industrial areas, Militia units could be raised in workplaces such as factories, mines or workshops. These units ranged in size from squads through to regiments and even divisions, depending on the size of the workplace. This badge was issued to Militiaman Zhang Heping（张和平）in a unit raised at the Shanghai New People's Machinery Factory, which together with others in the industry, formed the Militia Steel Regiment.

Urumqi Militia 乌鲁木齐民兵 c.1973.

Urumqi is the capital of the Xinjiang-Uyghur Autonomous Region of north western China, with a high population of Central Asian ethnic minorities. The badge of the Urumqi Militia features both Chinese characters and Pinyin, the system of Romanisation of the Chinese language. This was probably an aid for ethnic minorities whose first language was not Chinese.

Shanghai Militia 上海民兵，Shanghai Electrical Equipment Factory Battalion 上海电器工厂民兵营, c.1973

 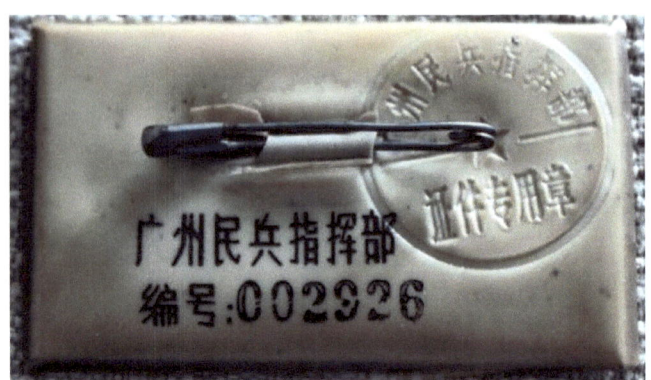

Guangzhou Militia 广州民兵, issued by the Guangzhou Militia Headquarters 广州民兵指挥部，1974.

In 1973, the radical leftists later known as the 'Gang of Four' attempted to use the Militia to build an armed power base in major cities. New 'urban militia' units comprised solely of workers were raised, along with new Militia headquarters solely under CPC control and independent of the PLA chain of command. The radicals of the CPC attempted to use their new militia as a counterbalance to their opponents, the political moderates who had the support of the PLA, in preparation for a power struggle after the impending death of Mao Zedong. Radical control of the urban militias only took root in Shanghai, the power base of the Gang of Four and the Shenyang Military Region. In other cities and in the countryside, the Militia remained firmly under dual CPC and PLA command in fact, if not in name (eg the Guangzhou Militia Headquarters). The Gang of Four prepared for an armed uprising by the urban militias in order to seize power upon the death of Mao Zedong, and an uprising actually took place in Shanghai in 1976, which was suppressed by the PLA.

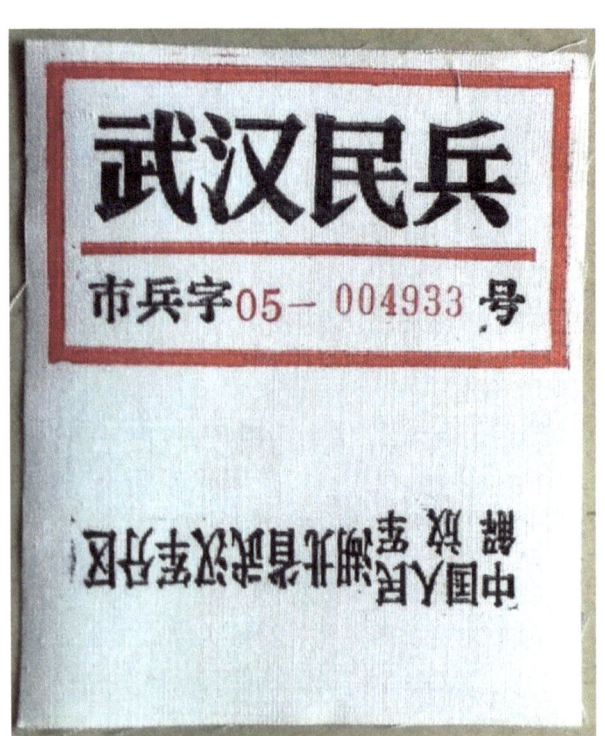

Wuhan Militia 武汉民兵，issued by the People's Liberation Army, Hubei Province, Wuhan Military Sub-District 中国人民解放军湖北省武汉军分区, c.1977.

This badge from the industrial city of Wuhan was issued by the PLA Wuhan Military Sub-District, the command responsible for the city's Militia units. This dates the badge probably to the period after the independent Militia Headquarters were abolished in 1977 and the formal reassertion of dual CPC-PLA command of the Militia.

# Militia in Print

*Illustrated Militia Training Manual*
(*Minbing Xunlian Tuce*), 1958

*Militia Pictorial*
(*Minbig Huace*), January 1959

*People's Pictorial*
(*Renmin Huabao*), June 1965

*China Reconstructs*
January 1970

# Militia in Posters

'Heroic Militiawoman' (*Yingxiong nu Minbing*) 1965

'Do Not Forget Class Bitterness. Keep A Firm Hold Of Your Rifle' (*Bu wang jiejiku. Jinwo shouzhong qiang*) 1965

Militia Certificate of Merit, similar to that shown in the 'Heroic Militiawoman' poster. Unissued dated 1970s

'Be Well Trained In Your Basic Skills. Protect The Motherland' (*Lianhao benling. Baohu zuguo*). Mei Xiaoqing c.1960

'Militia Military Training Poster No. 6' (*Minbing junshi xunliantu zhi liu*) 1958. Military and political training poster series, collated in the *Illustrated Militia Training Manual* (*Minbing Xunlian Tuce*), 1958.

'Accurate Shooting' (*Da de zhun*) 1965

Firing At Airborne Targets. (4. Firing At Enemy Paratroopers)
(*Dui kong sheji. 4. Sheji diren sanbing de fangfa*) 1970.

# Glossary

This is a glossary of common terms encountered by shooters and collectors of Chinese arms and accessories. Note that many of the various accessories and weapons in this book or others encountered by collectors were manufactured in the period from 1955 – 1965. This was a period of great change to the Chinese written script, as many 'traditional' characters were 'simplified' by reducing or changing the strokes required to write the character. Some words from modern technology or foreign languages were still being standardised, and variations featuring different characters were in use. Readers may note the variations in the characters or words shown on collector's items in this book, and the inconsistency in their use and in their forms. Most characters are listed in standard simplified characters, with the exception of where collector's items were only manufactured and labelled in traditional characters, or where traditional characters are shown for emphasis.

## Organisations

| | | |
|---|---|---|
| Militia | 中国民兵 | *Zhongguo Minbing* |
| People's Liberation Army | 中国人民解放军 | *Zhongguo Renmin Jiefangjun* |
| People's Armed Forces Department | 人民武装部 | *Renmin Wuzhuangbu* |
| Public Security Bureau | 公安部 | *Gonganbu* |

## Weapons

| | | |
|---|---|---|
| 30 Rifle (M1903 Springfield) | 三〇步枪 | *Sanling buqiang* |
| 65 Rifle (Type 38 Arisaka) | 六五步枪 | *Liuwu buqiang* |
| 79 Rifle (Mauser 7.92mm) | 七九步枪 | *Qijiu buqiang* |
| Type 50 Submachine Gun (PPSh-41) | 五０式冲锋枪 | *Wuling shi chongfengqiang* |
| Type 53 Carbine (Mosin Nagant) | 五三式步骑枪 | *Wusan shi buqiqiang* |
| Type 54 Submachine Gun (PPS-43) | 五四式冲锋枪 | *Wusi shi chongfengqiang* |
| Type 56 Semi-automatic Carbine (SKS) | 五六式半自动枪 | *Wuliu shi banzidongqiang* |
| Type 56 Assault Rifle (AK-47) | 五六式冲锋枪 | *Wuliu shi chongfengqiang* |
| Type 63 Automatic Rifle | 六三式自动枪 | *Liusan shi zidongqiang* |
| Type 99 Rifle (Type 99 Arisaka) | 九九式步枪 | *Jiujiu shi buqiang* |
| Assault Rifle | 冲锋枪 | *chongfengqiang* |
| Automatic Rifle | 自动枪 | *zidongqiang* |
| Carbine (bolt action) | 步骑枪 or 步马枪 | *buqiqiang* or *bumaqiang* |
| Pistol | 手枪 | *shouqiang* |
| Rifle (bolt action) | 步枪 | *buqiang* |
| Semi-automatic Carbine | 半自动枪 | *banzidongqiang* |
| Submachine Gun | 冲锋枪 | *chongfengqiang* |

## Ammunition Labels

Chinese ammunition cases, cans, paper packets and boxes were labelled in the standard Soviet labelling system beginning in 1956. The key ammunition data indicates the cartridge type, calibre, bullet type and case type. The bullet type is indicated in abbreviated form with one or two characters. The case type may be either brass or steel, which was introduced in 1956 and which replaced brass completely by 1965-66. It should be noted that lead core light ball (L) appears to have ceased production in 1956, with the introduction of the steel core light ball as the standard ball type for both Type 53 7.62x54mmR and Type 56 7.62x39mm cartridges. From 1956, steel core (LPS) was produced under the designation 7.62 钢轻 with either brass or steel cases until 1965, when it was redesignated 7.62 普 and produced exclusively with steel cases. Other 7.62x54mmR specialist cartridges and 7.62x25mm ball continued to be produced in brass cases until the supply was exhausted and then exclusively with steel cases from 1965-66. Apart from the early production light ball and the two designated 7.62 普 ball cartridges which appear only to have been produced in the specific bullet/case combinations listed, the cartridge labels listed below can be considered representative only, and cartridges may potentially be found in either brass or steel cases. The full glossary of ammunition label information is listed in the following section.

| | | |
|---|---|---|
| 7.62x54mmR Light Ball (L) [35] | 7.62 轻弹 | *7.62 qingdan* |
| Type 53 7.62x54mmR Ball (L) (brass) | 53 式 7.62 普（铜） | *53 shi 7.62 pu (tong)* |
| Type 53 7.62x54mmR Ball (LPS) (steel) | 53 式 7.62 普（铁） | *53 shi 7.62 pu (tie)* |
| Type 53 7.62x54mmR Steel Core Light Ball (LPS) (steel) | 53 式 7.62 钢轻（铁） | *53 shi 7.62 gang qing (tie)* |
| Type 53 7.62x54mmR Steel Core Light Ball (LPS) (brass) | 53 式 7.62 鋼轻（銅） | *53 shi 7.62 gang qing (tong)* |
| Type 53 7.62x54mmR Tracer (brass) | 53 式 7.62 曳（铜） | *53 shi 7.62 ye (tong)* |
| Type 53 7.62x54mmR Armour Piercing Incendiary (brass) | 53 式 7.62 穿燃（铜） | *53 shi 7.62 chuan ran (tong)* |
| Type 53 7.62x54mmR Incendiary (brass) | 53 式 7.62 试燃（铜） | *53 shi 7.62 shi ran (tong)* |
| Type 51 7.62x25mm Ball (steel) | 51 式 7.62 手普（钢） | *51 shi 7.62 shou pu (gang)* |

## Ammunition Information

| | | |
|---|---|---|
| Armour Piercing Incendiary (Soviet B-32) | 穿甲燃烧弹 | *chuanjia ranshaodan* |
| Ball (1956 Soviet L & post-1965 Soviet LPS) | 普通弹 | *putongdan* |
| Incendiary (Soviet ZP) | 试射燃烧弹 | *shishe ranshaodan* |
| Steel Core Light Ball (1956-1965 Soviet LPS) | 钢心轻弹 | *gangxin qingdan* |
| Tracer (Soviet T-46) | 曳光弹 | *yeguangdan* |
| Light Ball (1952-1955 Soviet L) | 轻弹 | *qingdan* |
| Ball (Pistol & Submachine Gun) (Soviet P) | 手冲普通弹 | *shouchong putongdan* |
| Brass | 铜 | *tong* |
| Charger | 夹 | *jia* |
| Steel (steel case rifle cartridge) | 铁 | *tie* |
| Steel (steel case pistol cartridge) [36] | 钢 | *gang* |
| Type | 式 | *shi* |
| Rounds | 发 | *fa* |

---

[35] Early production lead core light ball (L) with brass case 1952-1955. This early production ammunition was not labelled in the standard Soviet labelling system until 1956, when it was redesignated 53 式 7.62 普（铜）. The designation of the 7.62x54mmR cartridge as 'Type 53' indicates its use by a range of 'Type 53' weapons, ie Carbine, LMG and MMG. This system is paralleled with the Type 56 7.62x39mm cartridge, used by a range of 'Type 56' weapons, ie Carbine (SKS), Assault Rifle (AK-47) and LMG (RPD).

[36] The Chinese word for 'steel' is 钢铁 (*gangtie*) and the characters 钢 (*gang*) and 铁 (*tie*) are used as abbreviations in Chinese ammunition nomenclature. In Type 53 7.62x54mmR, 钢 indicates the steel core bullet, while 铁 indicates the steel cartridge case. In Type 51 7.62x25mm, 钢 indicates the steel cartridge case.

## Accessories and Field Equipment

| | | |
|---|---|---|
| Cartridge | 子弹 | *zidan* |
| Magazine (detachable) | 弹匣 | *danxia* |
| Pouch | 袋 | *dai* |
| Ammunition pouches (bandolier, chest pouches) | 子弹袋 | *zidandai* |
| Magazine pouch | 弹匣袋 | *danxiadai* |
| Sling | 背带 | *beidai* |
| Factory | 工厂 | *gongchang* |
| Year | 年 | *nian* |
| Year of manufacture | 年度 | *niandu* |
| Quarter of manufacture | 季度 | *jidu* |
| Manufactured | 制 | *zhi* |
| mm | 毫米 or 公厘 | *haomi* or *gongli* |
| Oil | 油 | *you* |
| Alkali (solvent) | 碱 | *jian* |
| Salted (solvent) | 咸 | *xian* |

## Common Traditional Characters

During the period from 1956 to 1960, inconsistencies in the use of simplified and traditional characters or standardised words were common. The characters in the following list are words commonly found in both traditional and simplified forms on the labelling of ammunition, accessories and field equipment. By 1960, it appears that all equipment and ammunition labelling had adopted the new simplified standard.

| | Traditional | Simplified | Pinyin |
|---|---|---|---|
| Brass | 銅 | 铜 | *tong* |
| Cartridge | 彈 | 弹 | *dan* |
| Charger | 夾 | 夹 | *jia* |
| Gun (SMG/Rifle/Pistol) | 搶 | 枪 | *qiang* |
| Manufactured | 製 | 制 | *zhi* |
| Rounds | 發 | 发 | *fa* |
| SMG/Assault Rifle | 衝鋒槍 | 冲锋枪 | *chongfengqiang* |
| Steel | 鋼 | 钢 | *gang* |
| Iron (steel case) | 鉄 | 铁 | *tie* |
| Light | 輕 or 轻 | 轻 | *qing* |

## Chinese Numerals

| 0 | 一 | 二 | 三 | 四 | 五 | 六 | 七 | 八 | 九 | 十 | 五0 | 五三 | 五四 | 五六 |
|---|---|---|---|---|---|---|---|---|---|---|---|---|---|---|
| 0 | 1 | 2 | 3 | 4 | 5 | 6 | 7 | 8 | 9 | 10 | 50 | '53 | 54 | 56 |

www.ingramcontent.com/pod-product-compliance
Lightning Source LLC
Chambersburg PA
CBHW041700160426
43191CB00002B/32